WITHDRAWN
Estes Valley Library

D0389883

FOR ENGINEERS AND SCIENTISTS

DAVID FISHER, EDITOR

Gulf Publishing Company
Houston, London, Paris, Zurich, Tokyo

This book was reviewed and revised by the editor, and reprinted January 1996.

RULES OF THUMB
FOR ENGINEERS AND SCIENTISTS

First Published in 1988 by Trans Tech Publications Ltd., Switzerland.

This revised and expanded edition published in 1991 by Gulf Publishing Company, Houston, Texas.

Library of Congress Cataloging-in-Publication Data
Fisher, D. J.
Rules of thumb for engineers and scientists / David Fisher, editor. — Rev. and expanded ed.
p. cm.
Includes bibliographical references and indexes.
ISBN 0-87201-786-9
1. Science—Handbooks, manuals, etc. 2. Engineering—Handbooks, manuals, etc. I. Title.
Q199.F57 1991
500—dc20 90-38439
CIP

Contents

* *See page xii for explanation of asterisks.*

C — 31

D — 49

E — 55

I — 93

K — 95

L — 107

M — 115

R — 141

S — 149

T — 169

Preface

In science, rules of thumb are the poor relations of laws and, although useful, cannot always be depended upon. Perhaps because of this, there tends to be a marked reluctance to disseminate them widely. At the same time, they are frequently proposed in the literature. The present compilation is an attempt to begin to bridge this gap between supply and demand. It should not be assumed to be an exhaustive list of all the rules of thumb that have been discovered. Rather, it should be regarded as a "sampler" of such rules and is a miscellany of those I have found particularly useful or surprising.

Thanks are due to the few who replied to the compiler's published appeals for rules of thumb. The aid of Professor F. Bénière, Professor A. A. Berezin, Dr. D. C. Robie, and Dr. A. M. Stoneham is hereby acknowledged.

David J. Fisher, D. Sc.

Introduction

I make no apology for the fact that this is a rather idiosyncratic reference work. It originated from a need to know the properties of an unusual class of organic compounds known as "plastic" crystals. It quickly became apparent that, although the properties important to organic chemists (melting point, boiling point) had invariably been determined, those important to a metallurgist (heat and solute diffusivities in the liquid and solid state, surface energies) had not. I was eventually forced to measure most of the required properties myself but, while searching fruitlessly for such data in *Chemical Abstracts* and other sources, I discovered many intimate correlations that existed between different properties and different substances. A particularly inspiring influence at this time was a paper by K. G. McNeill (*American Journal of Physics*, 1960, **28**, 375), in which the author predicts the properties of copper and lead on the basis of a few simple relationships. A later inspiration was E. M. Purcell, of the *American Journal of Physics*, and his "Back of the Envelope" column. For the record, one work that did *not* inspire the present one was the book *Rules of Thumb*, by T. Parker (Houghton Mifflin). Preparation of the present book was well under way when the latter work appeared. No doubt it is a rule of thumb in itself that when one finds a book title that appears never to have been used before, someone else has the same idea and gets to use it first.

Chance comments by colleagues also revealed that even "well-known" approximate correlations were not, in fact, well known. This lack of dissemination appeared to arise mainly from the false assumption that everyone else already knew them. The latter fact immediately became clear when, after deciding to collect as many newly proposed correlations as possible, I became aware of the many "classic" correlations of which I had hitherto been ignorant.

Consequently, this first compilation is a mixture of those correlations (identified by asterisks*) that I believe should become accepted as "rules of thumb," and those correlations that have already been acclaimed as rules of thumb by others. It is not easy to find the "classic" rules of thumb, and the present list is certainly incomplete. The difficulty stems from the fact that very few authors index the rules under "rules of thumb" or even under rules, again probably assuming that the reader already knows the name of the rule. Moreover, normally trusty sources such as *Chemical Abstracts* and *Science Citation Index* make a point of *not* indexing the word "rule" or indeed any likely synonym such as "correlation," "relationship," etc.

The question arises as to what constitutes a rule of thumb. I gave a great deal of thought to this problem and developed many sets of guidelines that limited how complicated a rule was allowed to be and what subjects it should cover. Unfortunately, adhering to any set of guidelines would have necessitated missing out on one or another particularly interesting rule. Finally, it was decided simply to include anything I felt like including—hence the idiosyncrasy.

One reason why the rule of thumb is less popular in science than in certain other professions may be that scientists are less likely to have to "think on their feet." On the other hand, the medical world is particularly rife with heuristics, so much so that when a doctor diagnoses, with great profundity, that a patient is suffering from gallstones, one cannot be sure whether this is the fruit of much weighing of the physiological evidence or the simple application of the rule "fat-fertile-female-fifty," as a handy guide to the probable incidence of the condition (*New Scientist*, 21st March, 1985). When judging the "correct weight" for a given height, a medic may simply subtract 100 from the patient's height (in centimeters) to give the permissible number of kilograms. If he is feeling particularly dedicated, he may use the "ponderal index," which is the weight (in kilograms) divided by the square of the height (in meters). The resultant value (which appropriately has the units of pressure) should be between 20 and 25. It is also reported that students of forensic medicine were once routinely vouchsafed the advice that, "if the number of bullet holes in a patient is odd, that patient has an odd number of bullets in him" (*British Medical Journal*, volume 292, p. 1399). Rather more cynical is the rule of thumb attributed to pharmaceutical manufacturers that, "there must be 10^5 victims of a disease before it becomes profitable to market a drug to treat it" (*Scientific American*, January, 1983, p. 54).

More seriously, there are several reasons why the collection of rules of thumb should be useful:

1. The first is the previously mentioned ability to "think on one's feet." The universal use of computers has already produced students who are rather out of touch with reality. If one asks such a student what the order of magnitude of a given quantity is likely to be, the result will probably not be an educated guess but a request to visit the ... computer terminal. This is not because the ability to perform simple calculations has been lost, but rather because the simulation mania has percolated downwards to the student level. Unfortunately, we live in a world that is increasingly ruled by the media and by mammon; one in which the

smart answer is a *sine qua non* of credibility. The person who can deliver the "technical fix" on the spot as well as do the real work later will get the attention or the funding.

2. Purcell, an editor of the *American Journal of Physics*, recognized this insidious loss of what might be termed "physical intuition" some years ago and started his "Back of the Envelope" column, which seeks to inculcate this lost art of approximation. One aim of *Rules of Thumb* is to perform the same service, although it does not duplicate the efforts of Purcell. Purcell's examples involve fitting approximate values into accepted physical laws. In contrast, the present work provides a wealth of sometimes surprising correlations whose accuracy can be questioned but into which one may as well fit the best data available and, again, obtain an order of magnitude estimate.
3. Recalling and naming rules can be useful, even if the principle is obvious, because this facilitates discussion and avoids a lot of hand-waving. See, for example, the Sidgwick-Powell rules, which "everyone" has been taught but few could name.
4. Having a rough estimate of how a material should behave can quickly eliminate anomalous results. See, for example, Sirdeshmukh and Subhadra (*Journal of Applied Physics*, 1986, **59**, 276) for a typical example.
5. Recalling a rule can help to avoid "re-inventing the wheel": a process that is increasingly wasting space in scientific journals, as pointed out by Dienes and Welch (*Physical Review Letters*, 1987, 59, 843).
6. Rules of thumb can help to maintain links between science and technology and avoid a rift appearing between them. As Van Uitert has stated (*Journal of Applied Physics*, 1981, **52**, 5547): "as the complexity ... increases, so does the likelihood that the technology-oriented reader will gain little from it. Hence, there are good reasons for presenting physical relations in a way that is easily understood and reducible to rules of thumb."
7. Although the use of computers was criticized, the computer programming industry is itself now a prime consumer of vague correlations. That is, the growing fields of "fuzzy logic" and "expert systems" are not averse to gathering together the vaguest of correlations and, nevertheless, obtaining valuable results.

However, there are some fields in which the use of rules of thumb is counterproductive. In mathematics, for example, it has been found that rules are no substitute for principles (T. H. Logan, *American Journal of Physics*, 1968, **36**, 79).

A

ABEGG'S RULE

The sum of the maximum positive valency exhibited by an element and of its maximum negative valency is 8.

A corollary is that the tendency to form compounds increases with increasing heteropolarity of the two elements involved. Nowadays, this is enshrined in the electronegativity concept. The above rule (Abegg, 1904) is generally true for groups 4 to 7, and is related to the 8-N and Octet rules (qv).

ADAMS' RULE

This useful rule serves to predict whether a given organic biphenyl compound (Figure 1) can be resolved. According to the rule:

A substituted biphenyl can be resolved if, and only if, the sum of the "hanging bond" lengths is greater than 0.29nm.

The definition of the "hanging bonds" is easily seen from the figure. The seemingly arbitrary value of 0.29nm is, in fact, the distance between the C atoms in the ortho position. Of course, the question of resolution (separation of racemates) also depends upon the temperature, experimental method, and other factors. Nevertheless, for simple compounds at temperatures of between 0° and 25°C, the rule is a useful guide. Basically, it works because it neglects the Van der Waals radii and the activation energy.

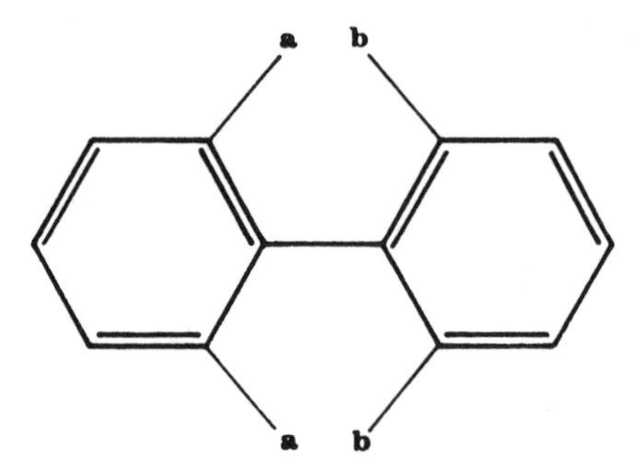

Figure 1. Adams' rule for the resolution of substituted biphenyls. The biphenyl can be resolved only if the sum of the C-a and C-b bond lengths is greater than 0.29nm.

ADJACENT CHARGE RULE

It is possible to write formal electronic structures for some molecules in which adjacent atoms have formal charges of the same sign. This rule states that such structures will not be important, due to instability resulting from the charge distribution. See also Pauling's rules.

AKHIEZER-DAVYDOV RULE*

Only those alloys which cluster can swell much less than the corresponding pure metals, and only those alloys which order can swell much more than the pure metals.

This rule (Akhiezer & Davydov, 1981) neatly summarizes the response of metallic alloys to the void swelling effect which is produced by neutron irradiation.

AL-BAYYATI RULE*

It is often desirable to have a short-cut method for choosing a sample size which is large enough to detect a significant difference between two sample groups or methods. Such a method was suggested (Al-Bayyati, 1971) for finding the sample size which is necessary to detect a difference in the true long-run proportions of two groups. It is assumed that nothing is known about the latter proportions; save that one wishes their difference to have a certain value. On this basis, it can be stated that:

The sample size necessary to give an 80% probability of detecting a difference, d, in the long-run proportions of two groups, at a significance level of 5%, is proportional to $(1 - d^2)/d^2$.

The proportionality constant is equal to 3.11 for a one-sided test, and is equal to 4.03 for a two-sided test (Table 1). Demanding a 95% probability of detection would almost double the sample size.

Table 1
The Al-Bayyati Rule
(Sample sizes required to detect difference between two samples; 95% confidence limits, 80% probability of success)

Difference (%)[1]	Sample Size[2]
5	1239
10	308
15	135
20	75
25	47
30	31
40	16
50	9
60	6
70	3
80	2
90	1

[1] *Difference in the percentages for each sample.*
[2] *One-sided test.*

ALDER RULE

In reactions of organic compounds such as cyclopentadiene, the substituted group can conceivably be found either outside of the cage formed by the other atoms (exo product), or inside the cage (endo product). These alternatives are shown in Figure 2 for the reaction of cyclopentadiene with maleic anhydride. According to the present rule:

In Diels-Alder reactions involving cyclic systems, addition is predominantly endo.

The rule can be rationalized on the basis that it gives a maximum overlap of the π electrons in the transition state, but may not always be valid.

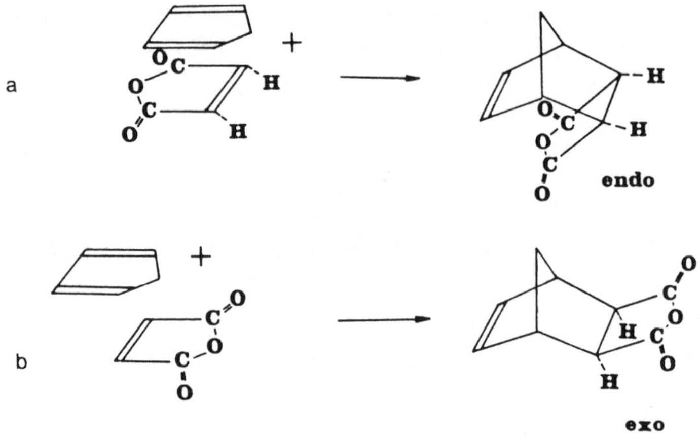

Figure 2. Illustrating the Alder rule. The Diels-Alder reaction of cyclopentadiene and maleic anhydride might occur so as to give either the endo (a) or the exo (b) form. The Alder rule predicts that only the endo form will be produced.

ALTONA-HAASNOOT RULE*

Like Benson's rule (qv), and several others in this compilation which permit the properties of a complex molecule to be deduced from the properties of its constituents, the present rule (Altona & Haasnoot, 1980) is an additive one. It states that:

> The proton-proton nuclear magnetic resonance coupling constants of pyranose rings in carbohydrates are given by the sum of the coupling contributions of its constituents.

Although a table of the group and topological contributions is required before one can make use of the rule, it significantly increases the ability to predict accurately the value of the proton-proton coupling constant of a 6-membered pyranose ring. Likewise, the probability of deducing the stereochemical characteristics of a new product from its measured coupling constants is considerably increased.

AMAGAT-LEDUC RULE

The volume occupied by a mixture of gases is equal to the sum of the volumes which the constituent gases would separately occupy at the same temperature and pressure as that of the mixture.

In the case of a perfect gas, this is the same as Dalton's law of partial pressures.

AMIDE RULE

See Hudson's rules.

ANDERSON'S RULE

When fabricating layer structures from semiconductors, it is necessary to satisfy electronic constraints at the interfaces. These include transport across the heterointerface, the sub-band energy of quantum-well structures, and band discontinuity. A useful guide to treating these problems is the electron affinity rule (Anderson, 1962), which states that:

The conduction band discontinuity at an interface between two different semiconductors is equal to the difference in the electron affinities of the adjacent semiconductors.

The valence band discontinuity is equal to the discrepancy between the energy gap difference and the conduction level difference. See also Bastard's rule and Dingle's rule.

ANTI-OCTANT RULE

See the Octant rule.

ANTONCIK'S RULE*

The effective use of ion implantation in semiconductor technology depends in part upon an understanding of the lattice location of the impurities which are introduced, and of their electrical properties. The general rule has been proposed (Antoncik, 1986) that:

Implanted atoms strive to be incorporated into the disordered lattice according to their natural valency unless they are forced, by the topological constraints of the host crystal, to occupy a substitutional site.

ANTONOFF'S RULE

$$\sigma_{12} = \sigma_1 - \sigma_2$$

That is: The interfacial surface tension between two saturated liquid layers is equal to the difference in the surface tensions of the two mutually saturated liquids against their common vapor.

The interfacial tension decreases with increasing temperature. Antonoff (1907) used to complain that the meaning of his rule was persistently misunderstood. Indeed, there is plenty of scope for confusion if one does not specify the exact meanings of the terms involved. Two corollaries of this rule are that the interfacial tension is less than the larger of the individual surface tensions, and that the equilibrium spreading coefficient ($\sigma_1 - \sigma_2 - \sigma_{12}$) of two partially miscible liquids is zero when the contact angle is zero. The rule fails in cases where the initial spreading coefficient of the liquid pair is negative.

AREA RULE

In aircraft designed for transonic flight, minimal drag is obtained by ensuring that the change in the cross-sectional area, from nose to tail, is smooth.

Since the presence of the wings leads to an unavoidable "bump" (Figure 3) in the cross-sectional area, this has to be compensated for (according to the above rule) by giving the aircraft a waist. At Mach numbers of between 1 and 1.05, 90% of the drag rise can be eliminated by using the rule (Streeter, 1962). At higher Mach numbers, the drag which results from using the area rule approaches the drag on a conventional aircraft. Wasp-waisted aircraft do not seem to be as common as they once were; probably because very few now need to operate in the transonic range.

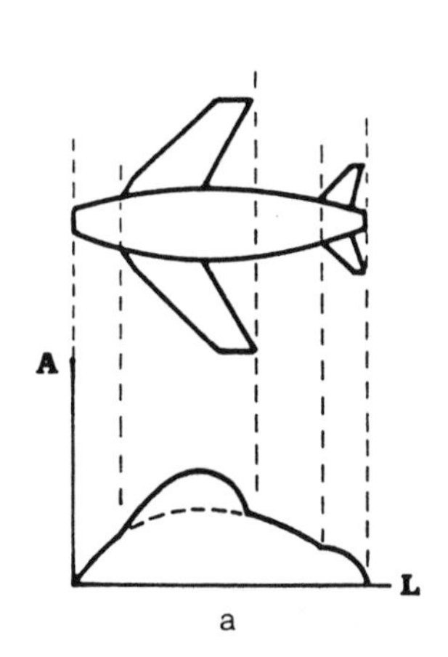

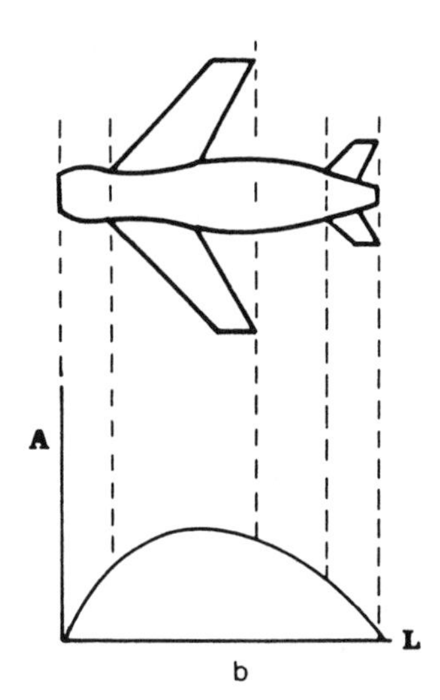

Figure 3. Illustrating the area rule. The unavoidable presence of wings leads to a "bump" in the plot of cross-sectional area versus aircraft length (a). According to the area rule, the resultant drag during transonic flight is minimized by smoothing the area/length plot. This can only be done by giving the aircraft a waist (b).

ARGILÉS' RULE*

Fischer projection formulae are widely used to represent compounds containing chiral centers. In these formulae, the horizontal bonds are assumed to extend forward from the plane of the paper, and the vertical bonds are assumed to be towards the rear. However, sugars often form hemiacetals. This involves the addition of an hydroxyl group to the carboxyl group, leading to a cyclic configuration of the molecule and the appearance of a new asymmetric C. This leads to alpha and beta anomers of each D- or L-type sugar. Such hemiacetals are more realistically represented by using the Haworth convention. Here, the heavy edge of the polygon is thought of as projecting outwards and the other edges are thought of as projecting into the paper. Although Haworth diagrams are easier to visualize, they are less easy to remember. Therefore, a rule was proposed (Argilés, 1986) for quickly converting from Fischer to Haworth representations (Figure 4). This says:

> Join the anomeric C atom to the other C atom in the hemiacetalic bond, using a D-shaped line for D-sugars or an S-shaped line for L-sugars. All OH-containing groups which fall inside the letter then project downwards in the Haworth structure, and OH-containing groups which fall outside of the letter project upwards in the Haworth structure.

Of course, the use of S for L-sugars is not arbitrary. It stands for "sinister"; that is, left-handed.

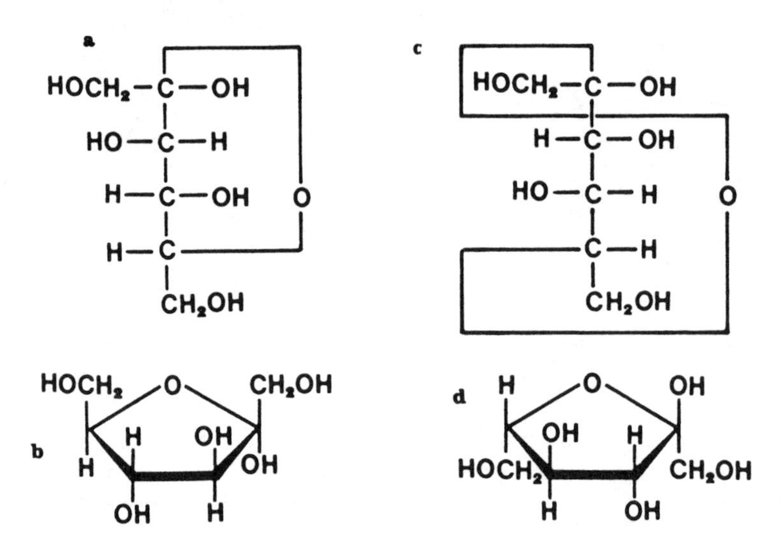

Figure 4. The use of Argilés' rule to convert Fischer diagrams into Haworth diagrams in the case of hemiacetals. In the case of alpha-D-fructose (a), a "D" is used to join the two C atoms of the hemiacetalic bond. The OH-containing groups which are inside (outside) the "D" are downward (upward) pointing in the Haworth diagram (b). In the case of alpha-L-fructose (c), an "S" is used to join the C atoms of the hemiacetalic bond. The OH-containing groups which are inside (outside) the "S" are downward (upward) pointing in the Haworth diagram (d).

ARITHMETIC COMBINING RULE

Combining rules are often used (Stwalley, 1971) in order to estimate the interaction potential, P_{AB}, of asymmetric pairs of atoms or molecules when those of the symmetric pairs (P_{AA}, P_{BB}) are known. One of the most commonly used is the rule that:

> The asymmetric pair interaction potential between two atoms or ions is given by:
>
> $$P_{AB}(R) = [P_{AA}(R) + P_{BB}(R)]/2$$
>
> where $P_{AA}(R)$ = A-A interaction potential at separation, R, etc.

This rule suffers from a disadvantage when the difference in the sizes of A and B is large. This is because the interactions in all three systems are assumed to occur at equal values of R. In order to avoid this problem, Smith (qv) and Lee & Kim (qv) proposed combining rules, for the repulsive interaction, which take account of this size difference. When an arithmetic mean value is used to estimate the effect of molecular size on surface tension, it is called the Lorentz rule. See also the geometric combining rule.

ARRHENIUS RULE

An often repeated rule, which is usually attributed to the Swedish chemist, states that:

> The rate of a chemical reaction approximately doubles with each 10°C rise in temperature.

ASCARELLI'S RULE*

It has been reported (Ascarelli, 1971) that the erosion resistance of pure metals is closely related to a parameter which he calls the "thermal pressure." This leads to the rule that:

> The erosion resistance of a metal is proportional to the product of the coefficient of linear expansion, the bulk modulus, and the difference between the melting point and the test temperature.

See also the rules of Brauer & Kriegel, Finnie-Wolak-Kabil, Hutchings, Khruschov, Smeltzer, and Vijh.

ASTON RULE

> The atomic weights of isotopes are approximately integers, and deviations of the atomic weights of the elements from integer values are due to the presence of several isotopes having differing weights.

See also Harkin's Rule.

AUWERS-SKITA RULE

This rule summarizes the relationship between certain physical properties and the conformation of an organic molecule. It was originally stated (Von Auwers, 1920; Skita, 1923) in the form:

> In general, cis compounds have a higher density and a higher refractive index than do the trans isomerides.

A more modern and extended form of the rule is that:

Among alicyclic epimers which do not differ in dipole moment, the isomer with the highest heat content (enthalpy) has the higher density, index of refraction, and boiling point.

Table 2 gives some typical data. Known exceptions to the rule include the boiling points of the alkylcyclohexanols, and any isomers which differ markedly in dipole moment. See the Van Arkel rule.

Table 2
The Auwers-Skita Rule
(Boiling point, refractive index, and density of dimethylcyclohexanes)

Isomer	Type	T_b(°C)	n	d
1,2	cis	129.7	1.4336	0.7922
1,2	trans	123.4	1.4247	0.7720
1,3	cis	120.1	1.4206	0.7620
1,3	trans	124.5	1.4284	0.7806
1,4	cis	124.3	1.4273	0.7787
1,4	trans	119.4	1.4185	0.7584

B

BABINET RULE

This very old rule (Babinet, 1838) suggests that:

In dichroic crystals, the faster ray is less absorbed.

However, there are numerous exceptions to the rule, and it fails for circularly dichroic media. See Bruhat's rule and Natanson's rule.

BADGER RULE

An empirical relationship (Badger, 1934, 1935) between the force constants and the vibrational frequencies of electron orbits in diatomic molecules. See also the Clark rule.

BANCROFT'S RULE*

In order to calculate the slope of the best straight line through a set of points, simply join the first and last points.

Many readers will throw up their hands in horror at this suggestion. Generations of students have been upbraided for doing just this; before being introduced to the method of least squares. The latter method is undoubtedly indispensable if one wishes to prove that a set of points obeys a linear physical law. However, Bancroft (1981) proposed a slightly different method and pointed out that, in the limiting case where one only wishes to know the gradient, one may as well use just the first and last points. In order to increase accuracy, one then has simply to increase the number of tests which are performed at the extreme ends of the range of the independent variable. In fact, Bancroft's method is rather more sophisticated than the extreme case considered here would suggest, and involves averaging the slopes between a number of pairs of points (Figure 5).

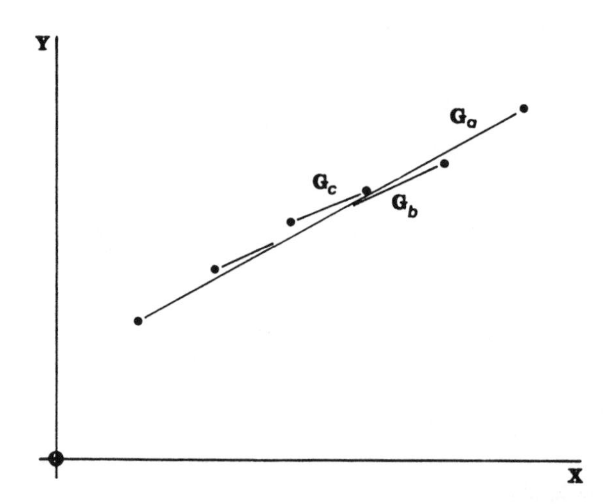

Figure 5. Illustrating Bancroft's rule. According to his method, an alternative to the least squares technique for treating linear data is to measure the gradients (G_a, G_b, G_c) between nested pairs of points and then to take a weighted average. In the present example, this weighted average is of the form:

$$(25G_a + 9G_b + G_c)/35$$

Because of the much greater weight given to G_a, a useful first approximation to the slope is obtained merely by joining the first and last points.

BANGWEI'S RULE*

One method of rationalizing the characteristics of compounds or alloys is to plot them onto graphs whose coordinates are various arbitrary functions of the fundamental properties of the constituents. One such "map" of this type is that of Mooser and Pearson (qv). Another type is that of Miedema. Although such maps often succeed in clearly demarcating different structures or behaviors, their disadvantage is that the line of demarcation usually has no obvious theoretical meaning or simple empirical representation. However, in the present case (Bangwei, 1983) it was found possible to distinguish between glass-forming and non glass-forming binary metallic alloys by means of a simple straight line drawn on the Miedema plot. The rule is thus that:

Binary metallic alloys form a metallic glass when W is greater than $3.9B^{1/3} - 0.1$.

Here, W is the difference in the work functions of the alloy components and B can be deduced from the difference in their bulk moduli.

BARAT-LICHTEN RULE

There are a large number of so-called "sum" and "correlation" rules which relate the energy levels in two separate atoms to those in a molecule. Most of these cannot be considered rules of thumb because they are either a) complicated in themselves or b) are simple, but link complicated integral functions of the atoms and molecules. Among the simplest rules is the present one (Barat & Lichten, 1972), which states that:

$$n_{u,r} = n_{i,r}$$

where $n_{u,r}$ and $n_{i,r}$ are the number of zeros of the radial wave function for the united atom and the isolated atom, respectively.

Because the rule leads to correlation diagrams which do not contain many quasi-crossings of the molecular orbital, it may be better to use the Eichler-Wille rule (qv). The use of this empirical rule in the construction of molecular orbital correlation diagrams calls for additional assumptions since the rule itself is satisfied by an infinite set of integrals. See also the Kereselidze-Kikiani rule.

BARBER'S RULE

A narrow ion beam issuing from one slit can be brought to a focus on a second slit, using a sector magnet, provided that the entry and exit trajectories are perpendicular to the magnet faces and the slits and the center of curvature of the mean beam are all co-linear.

This rule (Barber, 1932), which is more easily summed up by a diagram (Figure 6), assumes that the magnetic field is uniform. See also Cartan's rule for a more general case.

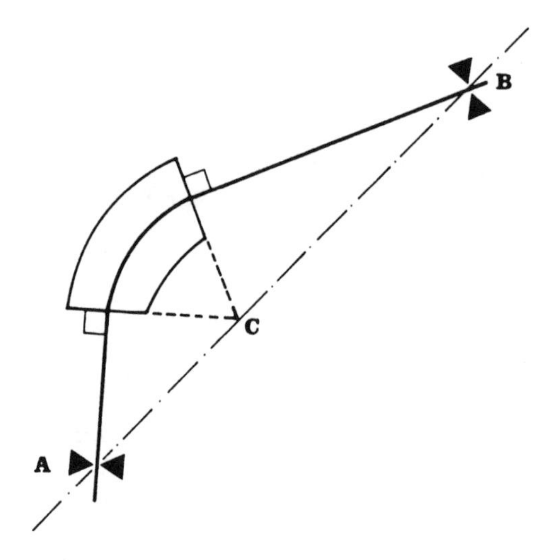

Figure 6. Illustrating Barber's rule for a sector magnet with a uniform field. A narrow ion beam from slit A can be brought to a focus at slit B, provided that the entry and exit trajectories are perpendicular to the magnet faces and that A, B, and the center of curvature of the mean beam, C, are all co-linear.

BARLOW'S RULE

The volume occupied by an atom in a molecule is proportional to the valency of the atom; where the lowest valency is assumed in the case of a multivalent element.

BARNSLEY'S RULE*

One of the most striking developments of recent times has been the explosion of interest in applying previously esoteric branches of mathematics to the real world. One of these branches has been the study of chaos—the non-deterministic behavior of solutions to equations which contain non-linearities. Closely related to this field is that of fractal theory. The latter recognizes that many natural systems exhibit patterns which are the same at all scales of magnification. Satellite photographs of entire deserts, for instance, can be indistinguishable from a photograph of the sand ripples which occur on a single dune. Increasingly often, practically important features such as fracture markings, slip patterns, and metallurgical microstructures, are described using the language of fractal theory. However, this use has invariably been descriptive rather than constructive; mainly because there appeared to be no "mathematical tool" available which would permit one to fit a fractal form to a "Chinese script" microstructure, say, in the same way in which one might fit a parabolic form to a growing dendrite. Such a tool now exists and is incredibly simple to use. It involves the application of just one rule, and is known to mathematicians as the "collage theorem" (Barnsley et al, 1986). The rule is:

> In order to obtain equations which reproduce a fractal object, "tile" the object with smaller copies of itself.

This obviously needs some elucidation. The main point is that this "tiling" is not an exact one, but a rather "slap-dash" one—hence the name, "collage". The technique is best appreciated by considering a concrete example. Suppose that one wishes to describe a leaf mathematically. The use of standard curve-fitting procedures would require a great deal of effort and would probably involve hundreds of parameters. Also, the function would have to be double-valued unless polar coordinates were used. Using the present rule, the leaf (Figure 7a) is photocopied at various reduced sizes (dotted lines in Figure 7b) and the copies are laid over the original. The gaps left by this "lazy tiling" could theoretically

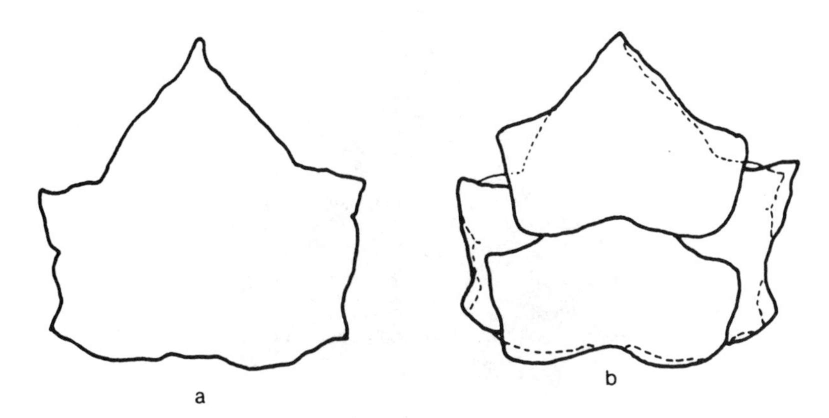

Figure 7. The use of Barnsley's rule (the collage theorem) to reproduce a leaf-shape. First step: the original form (a) is loosely covered with smaller copies of itself (b).

be filled by smaller and smaller copies. However, the four copies shown are quite sufficient. Having positioned them, one calculates the affine transformations which map the original leaf into the copies. In the present case, this gives four sets of simple linear equations:

(1) $u = 0.6x + 0.18$; $v = 0.6y + 0.36$
(2) $u = 0.6x + 0.18$; $v = 0.6y + 0.12$
(3) $u = 0.4x + 0.3y + 0.27$; $v = -0.3x + 0.4y + 0.36$
(4) $u = 0.4x - 0.3y + 0.27$; $v = 0.3x + 0.4y + 0.09$

where the (x,y) are the original coordinates and the (u,v) are the final coordinates, equations 1 and 2 describe the simple translations and reductions, and equations 3 and 4 describe the reductions, translations, and rotations. These, unbelievably, are sufficient to reproduce the form of the original leaf and, with practice, could be written down almost by inspection. In order to "recapture" the original shape, a rather unusual procedure is followed. One iterates the equations by inserting the (u,v) values back into the equations (i.e., setting x,y = u,v) and plotting the succession of points. HOWEVER, the new x,y values are not fed into the same equation, or into the next equation on the list (these alternatives soon lead to a fixed point). Instead, the next equation to be used is chosen at random. As this process is continued, a fractal approximation to the original leaf builds up on the computer screen (Figure 8).

Figure 8. The use of Barnsley's rule (the collage theorem) to reproduce a leaf-shape. Final step: the original form is recaptured using a series of iterated functions.

BARTENEV-SANDITOV RULE*

For glassy materials, ranging in type from oxides to polymers, it is found (Bartenev & Sanditov, 1982) that:

$$aT_gS/Y = \text{constant}$$

where a is the thermal expansion coefficient, T_g is the glass transition temperature, S is the fracture strength, and Y is the Young's modulus.

BASALO'S RULE

Using the SHAB rules (qv), guidelines can be formulated for the stabilization and isolation of metal complexes. The conditions of interest are those that best maximize the charge-control term. The rule which results is that:

Solid salts separate from aqueous solution most easily for combinations of either a small cation and a small anion or a large cation and a large anion, especially when there is also an equal but opposite charge on the cation and anion.

BASTARD'S RULE*

A current trend in semiconductor technology is to tailor the semiconducting properties of a material by the fabrication of layer structures. However, it is necessary to satisfy electronic constraints at the interfaces. The factors to be considered include transport across the heterointerface, the sub-band energy of quantum-well structures, and band discontinuity. It was suggested (Bastard, 1981) that:

For conservation of the probability current at the interface between two different semiconductors, the first derivative of the envelope function should be divided by the effective mass of the carriers.

When using this criterion, a 60% rule rather than the 85% rule (qv) may be more appropriate. See also Anderson's rule, Dingle's rule, and the Furdyna-Kossut rule.

BAY RULE*

Many years ago, I wished to use organic materials in order to simulate the solidification of eutectic alloys such as cast iron. Being extremely suspicious of the toxic and carcinogenic properties of organic materials,

I chose naphthalene as being one suitable constituent; mainly on the dubious basis that I had already been extensively exposed to it in the form of mothballs. Also, standard textbooks stated that it was not carcinogenic. The latter fact seemed curious. Why should some polycyclic hydrocarbons be innocent while others were very guilty of being carcinogenic? An afternoon spent with an elementary organic chemistry textbook and a manual on the dangerous properties of laboratory chemicals suggested a vague correlation between carcinogenicity and the characteristic pattern of lines shown in Figure 9a. However, the making of such a correlation between this property and the form of a schematic diagram seemed more like divination than science, and the idea (apart from the sketch) was forgotten. Some 12 years later, a rule was suggested (Accounts of Chemical Research, 1985, *17,* 332) to the effect that:

> Carcinogenicity in polycyclic aromatic hydrocarbons is associated with the presence of a bay region.

The so-called bay region is defined in Figure 9b. Note the connection with the compiler's sketch. Nowadays, far from being divination, the topological approach to predicting the properties of organic compounds is a growing field (Rouvray, 1986). No doubt a mystic or pseudoscientist, rather than using the tools of topology (Wiener index, Randic index) would make the quicker "a posteriori" connection that the bay region is like the space which is partially enclosed by a crab's claws and the

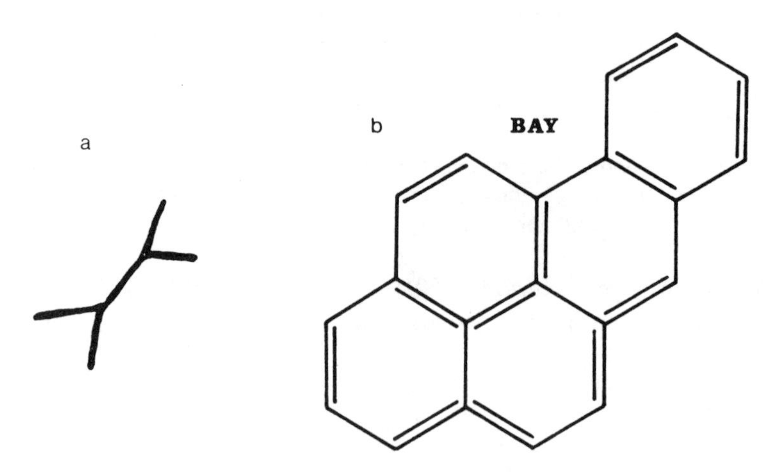

Figure 9. Defining the bay rule for carcinogenicity. The compiler's vague notion (a) and the current topologically-based bay criterion (b) of Herndon and Von Szentpaly. Note that molecular size considerations may over-ride the bay rule.

word, "cancer" comes ultimately from the word, "crab." The latter fanciful reflection is made merely in order to fix the rule in the reader's mind.

BEKE-KEDVES RULE*

By using the principle of corresponding states, useful rules can be found (Beke & Kedves, 1983) for diffusion in metals. These are that:

$$\ln [D_i/D] = 1.2 - 2.5(m/T)$$

for body-centered cubic metals, and

$$\ln [D_i/D] = 0.6 - 1.6(m/T)$$

for face-centered cubic metals, where D_i is the impurity diffusivity, D is the self-diffusivity, T is the temperature, and m is the initial slope of the median of the liquidus and solidus curves.

BENFORD'S RULE

Most scientists have followed standard introductory courses in statistics as applied to scientific fields. Such courses tend to treat only (a) "well-behaved" samples and (b) those parts of the normal probability distribution curve which are close to the crown of the "Admiral's hat." That is, zero bias in the data (apart from the effect one is studying) and completely random behavior of the errors are assumed. Sometimes, these conditions do not apply, and it is wise to be aware of the various pitfalls which exist. A case in point is the present rule (Benford, 1938). Unlike the others in this book, this rule does not offer a solution, but is a statistical oddity (thus making it part of the problem). Simply stated, Benford's rule says that:

There are more small things in the world than large things.

Obviously, if samples are chosen and analyzed in a manner which takes no account of this rule then very dubious conclusions may be reached. The rule is likely to apply when the property studied is, in some sense, a "conserved" one. Thus, the ages of the members of a population will not be subject to the rule because the aging of one person cannot be balanced by the rejuvenation of a second. On the other hand, masses or volumes or surface areas may well obey the rule. The effect of the rule is most striking when it crops up in everyday situations. For instance, Benford first detected the phenomenon in data on the surface areas of rivers. Rather than being randomly distributed, there were more 1's as first digits than there were 2's, 3's, etc. The effect was checked by using some 20,000 pieces of data, such as the addresses of randomly chosen scientists, and the distribution was found to be of the form:

$$P(n) = \log(n + 1) - \log(n)$$

where P(n) is the probability that n is the first digit, and the logarithms are taken to base 10.

Thus, there is a 30% probability that the reader's house number begins with the digit, 1. See also the rank size rule.

BÉNIÈRE'S RULE*

This was originally deduced (Bénière, 1974) from data on alkali halides, and states that:

$$S_s/S_f = \text{constant}$$

where S_s is the entropy of formation of a Schottky defect, and S_f is the entropy of fusion.

The Schottky defect here consists of a cation vacancy and an anion vacancy. A theoretical analysis indicated that the value of the constant was 4. However, experimental results suggest that its value is 3.2. Although derived for alkali halides, the rule may have wider applicability.

BENSON'S ADDITIVITY RULE

This rule (Benson, 1976) is simply a statement of the concept that:

The heats of formation of organic compounds are additive over their constitutive groups.

That is, the heat of formation can be found by adding individual contributions for each of the functional groups in the molecule. Thus, the heat of formation of 2,2-dimethylpropane (which looks like methane, but with all of the H atoms replaced by methyl groups) is found by summing one contribution for a C atom linked tetrahedrally to four C atoms, plus 4 contributions for a C atom linked to a methyl group. Evidently, the general use of the rule requires an extensive table (Carpenter, 1984) of group contributions. In practice, knowledge of just a few common group values suffices to estimate data for a wide range of compounds. In particular, it is very easy to predict stability (and therefore reactivity) trends in a homologous series. The difficulty is that contributions often arise from topological factors as well. Factors such as aromaticity and tautomerism must be taken into account, and can outweigh the functional group contributions.

BEREZIN'S RULES*

A number of rules, analogous to the Mollwo-Ivey rule (qv), have been proposed. One (Berezin, 1972) relates the energy of singlet-triplet splitting to the lattice constant:

$$[\text{splitting energy}]\,[\text{lattice constant}]^n = 1$$

Here, the exponent is equal to 2.3 for the M-center, and ranges from 0.5 to 0.9 for the F_t center. Another rule (Berezin, 1979) relates the binding energy of the F_{e+}' center to the lattice constant:

$$[\text{binding energy}] = A\,[\text{lattice constant}]^n$$

The constant and the exponent vary widely between different crystals. A third rule (Berezin, 1976) relates the two-photon annihilation time of the F_{e+}' center to the lattice constant:

$$[\text{annihilation time}] = A\,[\text{lattice constant}]^n$$

For alkali halides, A is equal to 0.226 and n is equal to 0.43 when the annihilation time is expressed in ns and the lattice constant is expressed in Angstrom units.

BEREZIN-VYATKIN RULE*

On the basis of measurements of the ionic conductivity of various superionic compounds having a disordered sub-lattice (Berezin & Vyatkin, 1986), it was suggested that:

$$Ca^2 = \text{constant}$$

where C is the ionic conductivity and a is the lattice parameter.

This relationship has been found to hold for materials with a body-centered cubic structure (Ag_2Se, Ag_2S, AgI), a face-centered cubic structure (Cu_2Se, Ag_2Te, CuAgSe), or a hexagonal structure (Cu_2Te, Cu_2S).

BERNAL-FOWLER RULES

So-called "ice rules" have been proposed (Bernal & Fowler, 1933) to the effect that:

1. The H atoms in ice lie on lines connecting the O atoms.
2. There is only one H atom between any given pair of O atoms.
3. Each O atom has two H atoms close to it and the unit of the water molecule is preserved.

Violations of the second rule lead to defects which are known as Bjerrum faults. The latter can be of two types. In the "doppelt" type, there are 2 protons between a pair of neighboring O atoms while, in the "leer" type, there are no protons between the O atoms.

BERTHELOT RULE

See Geometric Combining rule.

BINGHAM'S RULE

According to this old rule (Bingham, 1906), which is analogous to those of Richards (qv) and Trouton (qv),

$$ML_e/T_b = 17$$

where M is the molecular weight, L_e is the latent heat of evaporation, and T_b is the boiling point.

A more exact version was suggested to be $ML_e/T_b = 17 + 0.011T_b$.

BIOT'S RULE

This is also known as the additive rule for optical rotations, and is a typical example of a simple rule of mixtures:

$$a_m = xa_A + (1 - x)a_B$$

where a_m, a_A, and a_B are the specific rotations, at a given wavelength, of the mixture, of component A, and of component B, respectively, and x is the fraction of component A.

BIRGE-MECKE RULE

This is a semi-quantitative correlation between molecular constants of the various electronic states of a diatomic molecule. Thus:

$w/I = \text{constant}$

$wr^2 = \text{constant}$

where I is the moment of inertia, r is the equilibrium internuclear distance, and w is the equilibrium vibrational frequency.

BLANC RULE

This is a useful diagnostic rule for distinguishing between dibasic acids having a chain of 5 carbon atoms:

When a 5-membered ring is oxidized it gives glutaric acid and thence an anhydride, but when a 6-membered ring is oxidized it gives adipic acid and thence a ketone.

One exception is when a substituted compound of the adipic acid type gives a 7-membered cyclic anhydride, rather than a ketone.

BLANCK'S RULES*

By considering only the atomic numbers and mass numbers of the elements in the periodic table, it has been found to be possible (Blanck, 1989) to predict quite successfully the number of stable isotopes. The rules are:

1. Elements 1 to 7 each have 2 stable isotopes (except Be) with consecutive A values. The latter range from 1 to 15; omitting 5 and 8
2. Those elements from 8 to 83 which have an odd Z-value each have 1 or 2 stable isotopes. They all have odd A-values. (The A-values can usually be determined by rounding the molar mass to the nearest whole number.)
3. For elements 8 to 83, stable isotopes with even Z-values fill in the gaps which are left by the adjacent elements with odd Z-values.

Here, A is the mass number and Z is the atomic number.

BOGDAN'S RULE*

It was suggested (Bogdan, 1913) that the latent heat of evaporation of a liquid is related to other properties of the liquid by the rule:

$$L_eKd = 1$$

where L_e is the internal latent heat of evaporation, K is the compressibility of the liquid, and d is its density.

BOLDYREV-UVAROV'S RULE*

By analyzing literature data on ionic crystals (Uvarov et al, 1984), it was deduced that:

$$H/L = \text{constant}$$

where H is the activation enthalpy of charge carrier formation and L is the enthalpy of fusion.

The value of the constant was proposed to be 9.2. It was shown that the Barr-Lidiard relationship is a special case of this rule. It was also shown that the conductivity jump at the melting point can be deduced from the entropy of fusion. The present rule may be useful in predicting the conductivity of ionic crystals.

BOLLMANN'S RULE*

It is suggested (Bollmann, 1980) that the formation of Schottky defects in alkali halide crystals is governed by:

$$HN/L = 8$$

where H is the enthalpy of formation of the defects, N is Avogadro's number, and L is the latent heat of fusion.

BONZEL'S RULE*

$$E_s/T_m = \text{constant}$$

where E_s is the activation energy for diffusion on the surface of a metal, and T_m is the melting point.

This (Bonzel, 1972) is just one of a number of similar relationships between diffusion parameters and the melting point.

BOWDEN-JONES RULES*

A number of rules have been suggested for estimating the temperature dependence of various properties:

$L_e T = \text{constant}$

$dT^{3/5} = \text{constant}$

where L_e is the latent heat of evaporation, T is the temperature, and d is the density.

A better relationship in the case of water may be: $L_e T^{1/2} = \text{constant}$.

BOYER-BEAMAN RULE

$T_g/T_m = \text{constant}$

where T_g is the glass transition temperature and T_m is the melting point; both measured on the absolute temperature scale.

It is usually much easier to explain the glass transition temperature of a polymer once it has been measured than it is to predict the transition temperature from a knowledge of the structure alone. However, this rule (Boyer, 1954; Beaman, 1952) can be used to estimate the glass temperature for a polymer which crystallizes to some extent. The value of the above constant tends to lie between 0.5 and 0.75; suggesting that it is related (in a corresponding states sense) to the two-thirds rule (qv). More careful study shows that the ratio is usually close to 0.5 in the case of symmetrical polymers such as polyethylene (Table 3). The ratio is near to 0.7 in the case of asymmetrical polymers such as polystyrene and polyisoprene. The same rule appears to apply to metallic glasses (Voronel & Rabinovich, 1987).

Table 3
Boyer-Beaman Ratios of Selected Polymers

Polymer	T_g(K)	T_m(K)	T_g/T_m
polyethylene	163	408	0.40
polyvinylidene fluoride	234	483	0.49
polypropylene	255	449	0.57
Nylon 6	320	498	0.64
natural rubber (polyisoprene)	203	301	0.67
silicone rubber	150	215	0.70
polystyrene	373	503	0.74
polyvinyl chloride	355	453	0.78

BRAGG-KLEENAN RULE

The charged-particle stopping power of a material is proportional to the square root of the material's atomic weight, and is additive over the components of a complex molecule.

BRANCHING RULES

These rules are used in the construction of atomic term diagrams in order to obtain the multiplicities of the terms. For LS-coupling among the electrons, they state that:

1. For spin, the addition of an extra valence electron to a term of given multiplicity produces two terms having multiplicities which are one more and one less than the original.
2. For orbital angular momentum, the addition of an electron having orbital angular momentum, l, to a state having orbital angular momentum, L, gives states with total orbital angular momenta going from $|L - l|$ to $L + l$. By combining these results, the multiplicity of the new state can be found.

BRAUER-KRIEGEL RULE*

It has been reported (Brauer & Kriegel, 1965) for metals and alloys that:

There is a clear relationship between the elastic modulus and the resistance to erosion.

See also the rules of Ascarelli, Finnie-Wolak-Kabil, Hutchings, Khruschov, Smeltzer, and Vijh.

BREAKDOWN RULES

These help to rationalize the fragmentation behavior of organic molecules during mass spectrometry studies (Bloom et al., 1948). They state that:

1. The relative height of the parent peak is greatest for the straight-chain compound and decreases as the degree of branching increases.
2. The loss of a fragment containing a single C atom is unlikely unless the compound contains methyl side-chains.
3. Fragmentation is most likely at highly-branched C atoms.
4. Ions of odd mass tend to be more abundant than those of even mass, and secondary fragmentation involving the loss of H or larger fragments also tends to give ions of odd mass; especially in the case of straight-chain compounds. Most of the ions are formed by the fragmentation of a single C-C bond in the parent ion.
5. Prominent peaks at even mass numbers suggest the fragmentation of 2 separate side-chains, and imply a high degree of branching.
6. The study of metastable peaks often clarifies the steps leading to the formation of a particular ion.
7. In the spectra of paraffins, the peaks which correspond to C_3 and C_4 ions are always large.

Rule 3 is the most widely applicable one. See also Stevenson's rule and the rule of 13.

BREDT RULE

Reactions of Diels-Alder type with cyclopentadiene give cyclanes with a methylene bridge across a cyclohexane ring.

The corresponding alcohol does not hydrate because double bonds cannot exist in the bridge.

BREWSTER'S RULES

These are some of the many rules which have been deduced in order to relate the optical polarization behavior to the structure of an organic compound. The basic hypothesis (Brewster, 1959) used here is that a center of optical activity can be described as an asymmetric screw pattern of polarizability. Thus, in the case of the tetrahedral system, the rule proposed is that:

1. An asymmetric atom in the absolute configuration below is dextrorotatory when the polarizabilities of the substituent atoms decrease in the order: A - B - C - D.

```
      B
      |
  A — X — C
      |
      D
```

In the case of a twisted chain of 4 atoms, the rule is that:

2. A twisted chain of atoms, A-C-C-A′, of the form shown below is dextrorotatory for the second two configurations.

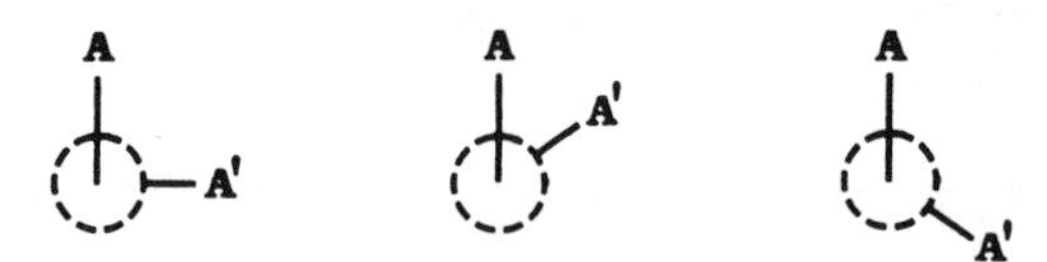

In addition to these rules of optical activity, which have since been extended to a wide range of configurations, the original paper above also suggested some rules of conformational analysis:

3. Only conformations which correspond to energy minima will be prevalent enough to produce appreciable rotatory effects.
4. All allowed configurations can be considered equally probable, to a first approximation.
5. The rotatory contributions of individual conformations are additive.
6. The 5-atom configuration below is prohibited when the terminal atoms are both larger than H.

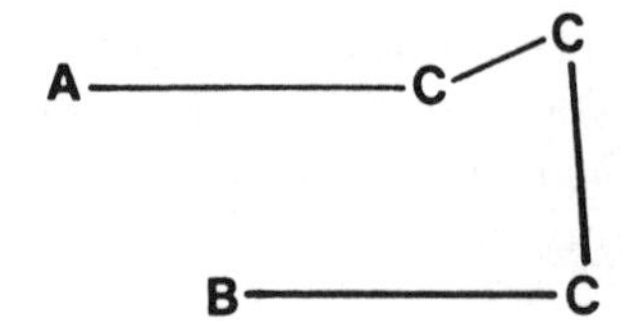

7. The configuration below is prohibited when the A, B, and C atoms are all larger than H.

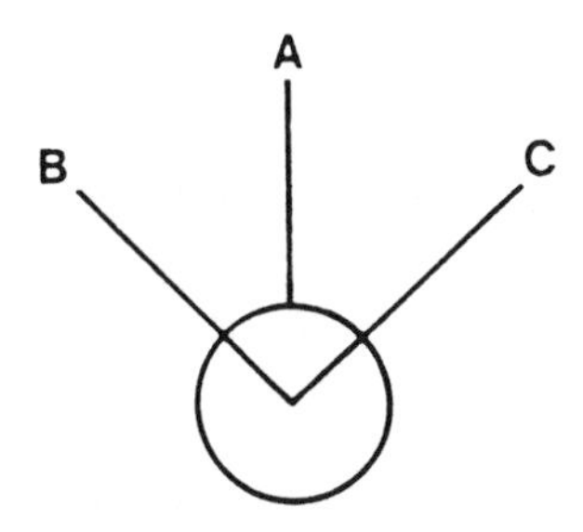

See also the Conformational rules.

BROWN'S RULE*

This (Brown, 1967) is a modification of one of Shewmon's rules (qv) and can be used to estimate diffusion parameters for an intermetallic solid solution which exhibits long-range order:

$$E = 36T_m + ZRT_c/4$$

where E is the activation energy for diffusion, Z is the coordination number, R is the gas constant, and T_c is the critical temperature for long-range ordering.

BRUHAT'S RULE

When a plane-polarized ray of light traverses an optically active medium, dextrorotation results if the right circularly polarized component is transmitted faster than the left one, and vice versa. The present rule (Bruhat, 1915) suggests that:

On the red side of an absorption band, the light ray which is less absorbed propagates with the greater velocity. The reverse is true on the violet side of the band.

BURDEN'S RULE

This rather old rule (Burden, 1871) suggests that:

$$T_b/M^{1/2} = \text{constant}$$

where T_b is the boiling point and M is the molecular weight.

Later work revealed numerous exceptions. However, it can be quite accurate when, for instance, members of a homologous series are compared.

BURGER-DORGELO-ORSTEIN RULE

For small multiplet splitting in atomic spectra where Russell-Saunders coupling applies, the sum of the intensities of all of the lines of a multiplet which belong to the same initial or final state is proportional to the statistical weight of $2J + 1$ of the initial or final state, respectively. Here, J is the quantum number for the total angular momentum of the electrons.

CARNELLY'S RULE

This classic rule (Carnelly, 1878) suggests that:

Elements with high melting points have low coefficients of thermal expansion.

This rule has some obvious exceptions (Table 4). However, in the case of Sn the failure of the rule may be related to the effect of the polymorphic transformations of Sn.

Table 4
Carnelly's Rule
(The coefficients of thermal expansion of the elements tend to decrease with increasing melting point.)

Element	Melting Point (C)	Coefficient (/°C)
Na	98	0.000075
Sn	232	0.000021
Pb	327	0.000029
Ag	961	0.000019
Cu	1083	0.000017
Fe	1527	0.000011
Pt	1774	0.000009

CARPENTER'S RULE*

The writing of the ground state electron configurations of the elements, according to the aufbau principle, is a standard exercise for students. The configurations are also the starting point for some of the other rules in this compilation. The applicability of some of Chapnik's rules (qv) depends upon the electron configuration. Various methods have been proposed for remembering these configurations. Carpenter (1983) proposed the arrangement shown in Figure 10; where one follows the arrows in order to read off the configuration. It has various drawbacks as a mnemonic, such as the "unnatural" method of reading the configuration. See also the Yi rule and the Hovland rule.

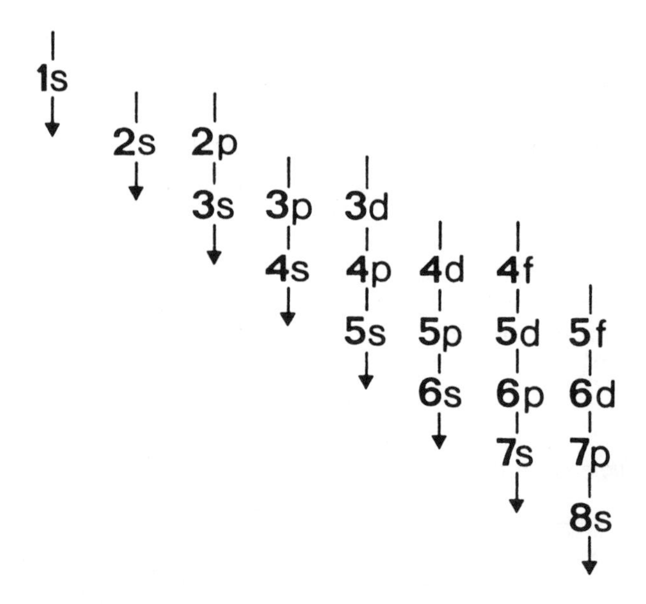

Figure 10. The Carpenter rule for writing the ground state electron configurations of the elements. The configurations are obtained by following the arrows.

CARTAN'S RULE

> A narrow ion beam issuing from one slit can be brought to a focus on a second slit, using a sector magnet, provided that the projections of the slits and perpendiculars to the magnet faces intersect at points which are co-linear with the center of curvature of the mean beam.

This rule (Cartan, 1937) is a generalization of Barber's rule (qv) in that it works even if the entry and exit trajectories are not perpendicular to the magnet faces. Like Barber's rule, it is more readily recalled in diagrammatic form (Figure 11).

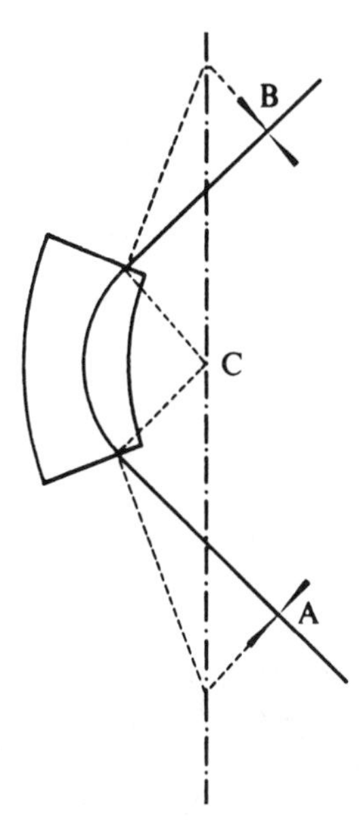

Figure 11. Illustrating Cartan's rule for a sector magnet with a uniform field. A narrow ion beam from slit A can be brought to a focus at slit B provided that the projections of the slits and perpendiculars to the magnet faces intersect at points which are co-linear with the center of curvature of the mean beam, C.

CASPER'S & VAN VEEN'S RULE*

While discussing the use of thermal helium desorption spectrometry to study the defect chemistry of metals (Casper & Van Veen, 1981), it was emphasized that the presence of a limited defect concentration was of utmost importance. This was expressed in the form of the rule that:

> In the study of defect chemistry using the THDS method, the defect concentration should be so low that the defects are closer to the crystal surface than to each other.

CATCHPOLE-HUGHES-INGOLD RULE

This rule (Catchpole et al, 1948) concerns the progress of organic reactions via unstable intermediates, and states that:

> When a proton is supplied by acids to the mesomeric anion of weakly ionising tautomers of markedly unequal stability, then the tautomer which is most quickly formed is the thermodynamically least stable and is also the tautomer from which the proton is lost most quickly to bases.

A typical example is that of pseudo-acids. Thus, when a salt of phenylnitromethane is acidified, the first-formed isomer is the thermodynamically unstable aci-nitro compound. However, the isomer which is ultimately formed is the thermodynamically stable nitro-compound. The aci-nitro compound much more quickly yields salts with bases than does the nitro-compound.

CHAPNIK'S RULES*

In a series of papers (Chapnik, 1980, 1984a, 1984b, 1984c, 1985, 1986), a number of comparative relationships were suggested between the semiconducting or superconducting properties of compounds and the positions of the components in the periodic table, or deviations in atomic volume of the compound from the ideal:

1. A compound in which the transition metal component is from the second long period, and has from 3 to 7 outer electrons, exhibits a higher superconducting transition temperature than that of an analogous compound in which the transition metal component is from the first or third period.
2. For the occurrence of superconductivity in a metal, the Hall coefficient which corresponds to the high-field region must be positive.
3. Semiconducting compounds of elements from the second period, which have from 4 to 9 outer electrons, have larger energy gaps than those of analogous compounds of elements from the first period. When the element has from 12 to 16 outer electrons, the opposite is true.
4. In the case of transition metal sulphides, metallic properties are associated with an atomic volume contraction which is greater than 21 and semiconducting properties are associated with an atomic volume contraction which is less than 21%.

Thus, in the case of the third rule, the energy gap (8eV) of ZrO_2 is greater than that (3eV) of TiO_2. On the other hand, the energy gap (1.9eV) of In_2S_3 is smaller than that (3.4eV) of Ga_2S_3. The volume contraction mentioned above (Table 5) is the percentage deviation of the atomic volume from the sum of its components; multiplied by a correction factor. This correction factor is simply equal to $(m + n)/2n$, where m and n are the numbers of atoms of the transition metal and of sulphur, respectively, in the compound. See also Matthias' rules.

Table 5
Chapnik's Rule for Transition Metal Sulphides.
(Semiconduction is associated with an atomic volume contraction which is less than 21%.)

Compound	Contraction (%)	Semiconductor
Re_2S_7	5	yes
VS_4	6	yes
Hf_2S_3	8	yes
PtS	10	yes
Hf_3S_4	11	yes
PdS_2	12	yes
ZrS_3	14	yes
TaS_3	15	yes
PtS_2	16	yes
Y_2S_3	17	yes
TiS_3	19	yes
LaS_2	21	yes
Ti_5S_8	21	no
Y_5S_7	23	no
Ti_8S_{12}	24	no
La_3S_4	25	no
Ta_2S	27	no
Ti_3S_4	28	no
Pd_3S	29	no
Ti_4S_5	30	no
Hf_2S	31	no
Y_4S_3	34	no
ZrS	35	no
TiS	37	no
NbS	39	no
$Nb_{21}S_8$	41	no

CHIRALITY RULE

See Wolf's rule.

CHO'S RULE*

It has been demonstrated (Cho, 1977) that the bulk modulus of a metal can be deduced from the rule:

$K/ne = \text{constant}$

where K is the bulk modulus, n is the number of valence electrons per unit volume of the solid metal, and e is the Fermi energy at absolute zero.

CHVORINOV'S RULE

In casting metals, the material in the riser serves to compensate for the contraction in volume which occurs during the solidification of that part of the casting which the riser is intended to feed. In order to ensure that this in fact happens, it is advisable to design the riser so that it remains molten longer than does the casting. Provided that the thermal environment and moulding material are the same for both riser and casting, the relative solidification rates of the latter can be estimated by using this rule, which states that:

The solidification time is proportional to the square of the volume of the metal, and is inversely proportional to the square of the surface area; that is:

$$\text{solidification time} = \text{constant} \times (\text{volume/surface area})^2$$

CLARK RULE

This relates the equilibrium internuclear distance of a diatomic molecule to the equilibrium vibrational frequency. It is based upon a broad classification of molecules and states that:

$$wr^3n^{1/2} = \text{constant}$$

where w is the equilibrium vibrational frequency, r is the equilibrium internuclear distance, and n is the group number.

The constant is made up of two components. One of these characterizes the period and the other is a correction for singly ionized molecules of that period. It is equal to zero for neutral molecules.

CLEMENS' RULE*

On the basis of a study (Clemens, 1986) of amorphous phase formation in sputter-deposited alloys, the rule was proposed that:

Amorphous phase formation by solid-state reaction is favored by a large driving force for reaction and by a low atomic size ratio of the smaller reacting atom to the larger one.

CLOUGH RULE

This (Clough, 1918) is one of the many rules which concern the relationship between molecular structure and the rotation of the plane of polarized light. It states that:

A mono-asymmetric alpha-hydroxy acid isomer which becomes more dextro-rotatory upon the addition of acid to an aqueous solution of its metal salt is of the L-configuration, whereas a shift of optical rotation in the laevorotatory direction is characteristic of the D-isomer.

CLOUGH-LUTZ-JIRGENSONS' RULE

This rule (Lutz & Jirgensons, 1930, 1931) is very similar to the previous one. However, they are usually treated quite separately in the literature:

If the molecular rotation of an optically active amino acid is shifted in a more positive direction upon adding acid to its aqueous solution, the amino acid has the L-configuration. A negative direction of shift is characteristic of a D-amino acid.

See Hudson's rules.

COEHN'S RULE

During electrification by friction, the material with the higher dielectric constant becomes positively charged to a density of $15(\epsilon_1 - \epsilon_2)$ C/m^2, where the ϵ are the permittivities of the two materials.

COLSON'S RULE *

It was suggested (Colson, 1887) that, for organic compounds:

H_m/T_m = constant

where H_m is the latent heat of fusion and T_m is the melting point in degrees Kelvin. When applied to metals it is known as Richard's rule (qv).

COMBINING RULES

See the arithmetic combining rule, the geometric combining rule, the Lee-Kim rule, and the Smith rule.

COMMON ANION RULE

It was first noted, when studying the Schottky barrier heights of semiconductors (McCaldin et al., 1976), that:

> In III-V and II-VI compounds, the position of the Fermi level relative to the valence band at the interface depends only upon the anion.

Compounds which contain aluminum are an exception to this rule.

CONFORMATIONAL RULES

There are a large number of conformational rules (Eliel et al., 1967; Hanack, 1965; Lister et al., 1978; Lemieux, 1971) for organic molecules. However, it is difficult to track most of them back to a specific originator and, rather than conferring eponymity on the wrong person, the compiler prefers to give these rules in the form of a "potpourri."

In five- or six-membered rings:

> A quaternary substituent always takes the equatorial position and vicinal substitution is disallowed. A tertiary substituent behaves like a quaternary one if, being axial, it is forced to point one of its non-H branches over the ring (Figure 12).

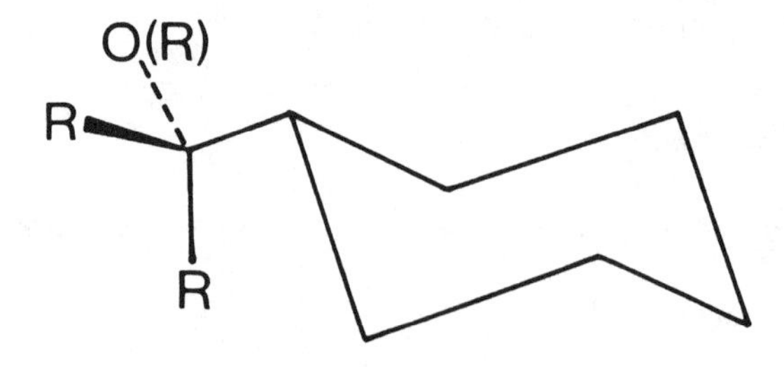

Figure 12. Conformational rules. A quaternary substituent always takes the equatorial position.

An axial lactol group is preferred over an equatorial one, and this orientation is encouraged by H-bond bridging via a syndiaxial hydroxyl group. The latter anomeric effect can be overridden if compensation is provided by some specific substitution pattern in the same ring (Figure 13).

Figure 13. Conformational rules. An axial lactol group is preferred over an equatorial one and is encouraged by H-bond bridging via a hydroxyl group.

Syndiaxial strain is avoided at all costs; except for H-bridged syndiaxial O atoms (Figure 14).

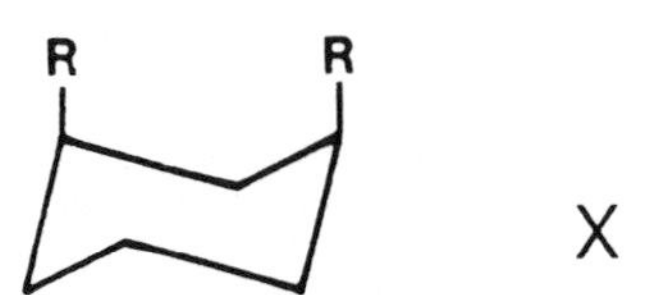

Figure 14. Conformational rules. Syndiaxial strain is avoided at all costs.

An equatorial tertiary group has only one favorable orientation when adjacent to one equatorial trans vicinal partner (Figure 15).

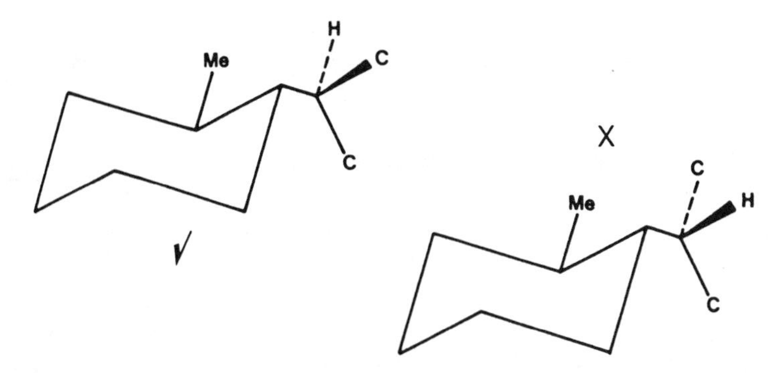

Figure 15. Conformational rules. An equatorial tertiary group has only one favorable orientation.

An equatorial secondary group has 2 allowed rotations when adjacent to one equatorial trans vicinal partner (Figure 16).

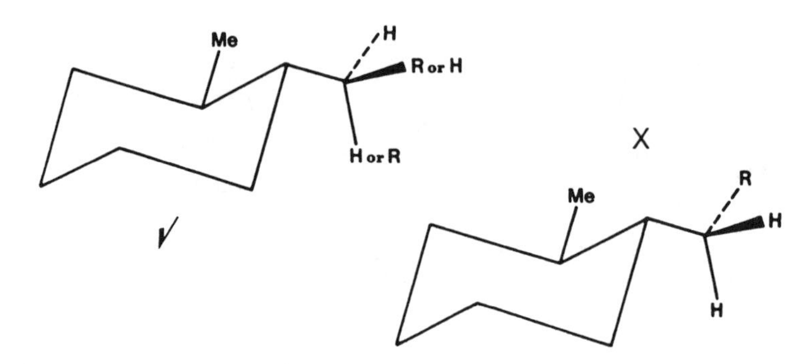

Figure 16. Conformational rules. An equatorial secondary group has 2 allowed rotations.

An axial tertiary group is made unfavorable by one equatorial cis vicinal partner. A second such partner aggravates the situation yet more (Figure 17).

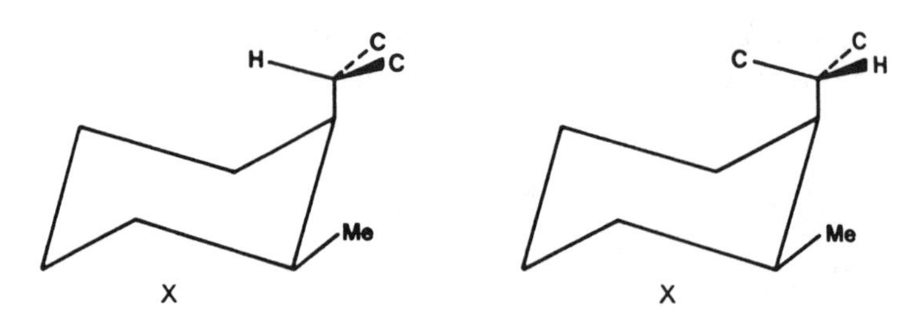

Figure 17. An axial tertiary group is made particularly unfavorable by a second equatorial cis vicinal partner.

An axial secondary group with an equatorial cis vicinal partner has one allowed rotational state. The tri-vicinal substitution pattern: equatorial-axial-equatorial, is thus possible only if the mid-axial group is a methyl one (Figure 18).

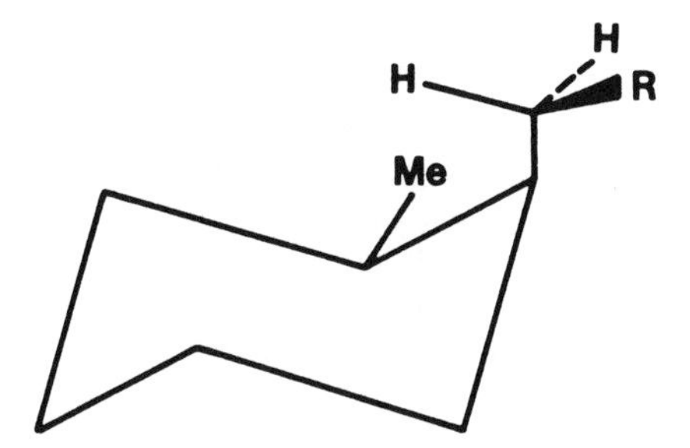

Figure 18. Conformational rules. An axial secondary group with an equatorial cis vicinal partner has one allowed rotational state.

An equatorial tertiary group has only one allowed rotational state with an axial cis vicinal partner. An additional axial flanking will not impair this preference, but an additional equatorial flanking is not allowed (Figure 19).

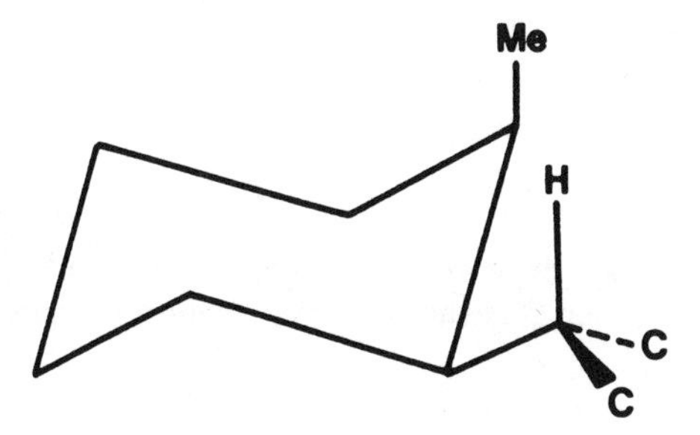

Figure 19. Conformational rules. An equatorial tertiary group has only one allowed rotational state with an axial cis vicinal partner.

An equatorial secondary group has 2 allowed rotations; even with an axial cis vicinal group. This is reduced to one state if it becomes flanked by one axial and one equatorial neighbor (Figure 20).

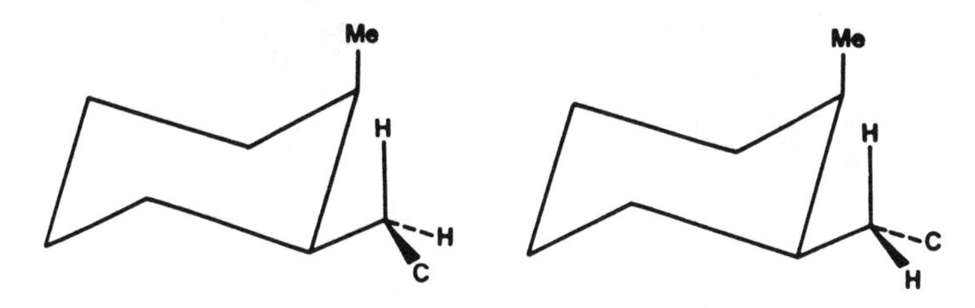

Figure 20. Conformational rules. An equatorial secondary group has 2 allowed rotations even with an axial cis vicinal group.

In acyclic rotors:

Two vicinal quaternary centers are allowed if they carry at least two O substituents. The preferred rotor is the one with maximum O-staggering but with a minimum of parallel oriented lone pair orbitals (Figure 21).

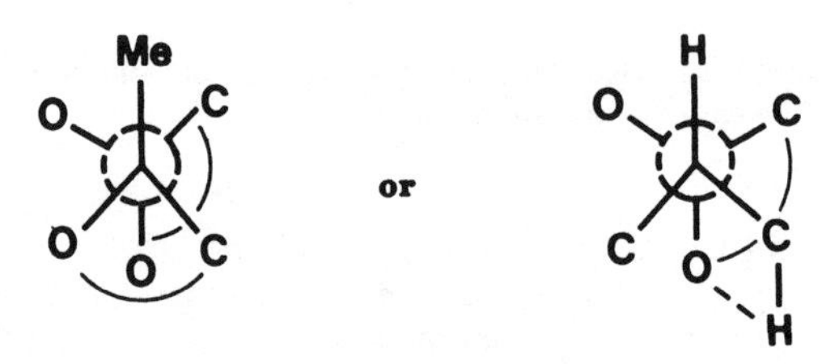

Figure 21. Conformational rules. Two vicinal quaternary centers are allowed if they carry at least two O substituents.

If the above rotor is in the connection between two rings, one of which is a tetrahydropyrane, then O-inwards rotation is preferred over C-inwards rotation (Figure 22).

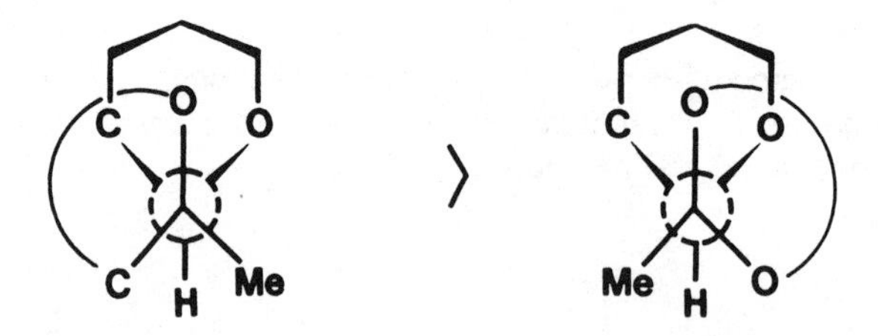

Figure 22. Conformational rules. Oxygen-inwards rotation is preferred between two rings.

The preferred rotor for a lactol implanted on a ring is the one with a minimum parallel orientation of the lone pair orbitals (Figure 23).

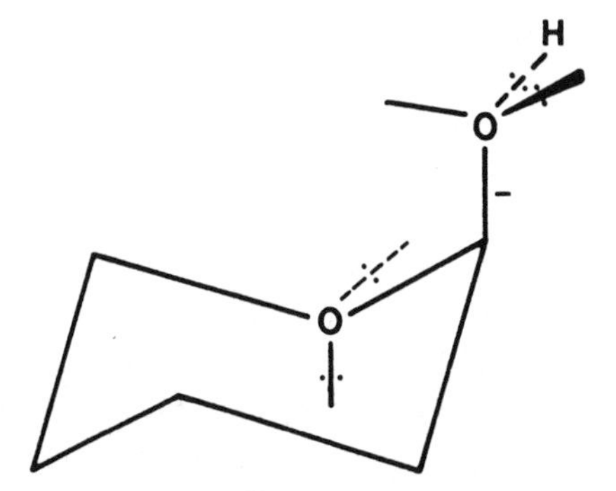

Figure 23. Conformational rules. The preferred rotor for a lactol on a ring has minimal parallel orientation of the lone pair orbitals.

Rotors possessing quasi-syndiaxial strain are disallowed, except when sp^2 hybridization is part of the quasi-syndiaxial path or a H-bridge mechanism can operate (Figure 24).

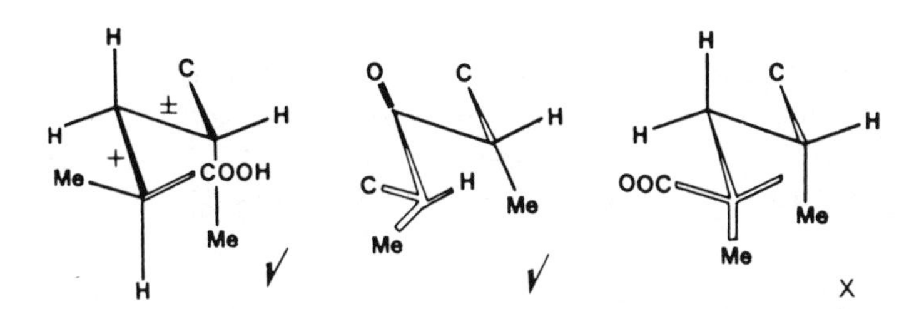

Figure 24. Conformational rules. Rotors possessing quasi-syndiaxial strain are disallowed.

A so-called "sandwich" position is slightly less preferred than 2 independent clinal dispositions (Figure 25).

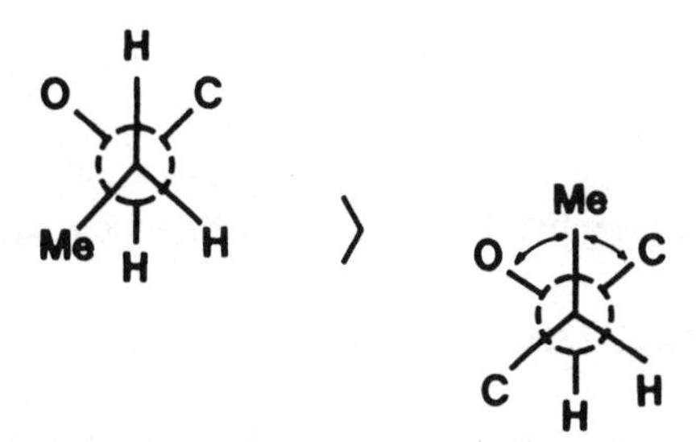

Figure 25. Conformational rules. The "sandwich" position is less preferable.

CONNECTION RULES

See Bastard's rule, Dingle's rule, the 60% rule, and the 85% rule.

COS² RULE

Nuclear magnetic resonance spectroscopy is one of the most powerful techniques which is available for the conformational analysis of heterocyclic organic compounds. The presence of electronegative heteroatoms gives heterocycles a much wider range in chemical shifts than is possible in alicyclic compounds, and increases the power of nuclear magnetic resonance techniques. In many cases, the ^1H spectra of heterocyclic compounds can be deduced from a first-order treatment of the coupling constants, and it is often assumed that:

$$J = A \cos^2 x$$

where J is the vicinal ^1H coupling constant, A is a constant, and x is the dihedral angle of a CH-CH fragment.

From this equation, it is immediately seen, for instance, that trans vicinal coupling (where the angle is equal to 180°) is greater than gauche vicinal coupling (where the angle is equal to 60°). The above expression is an approximation to the Karplus (1959) equation: $J = A\cos(2x) + B\cos(x) + C$.

COUCHMAN'S RULE*

Since both vacancy formation and the creation of new surfaces can be seen as being phenomenologically analogous, a relationship was sought (Couchman, 1975) between the two phenomena. This led to the rule that:

$$E_f/sr^2 = \text{constant}$$

where E_f is the monovacancy formation energy, s is the work required to create unit surface area of solid, and r is the relevant interplanar spacing. It was further estimated that the value of the above constant is about 2.8.

COULSON-RUSHBROOKE RULE

It can often be difficult to determine whether benzenoid hydrocarbons are of open-shell or closed-shell type. Perifusenes (peri-condensed benzenoid polycyclic systems) which have an odd number of C atoms and have no side-chains must be $(2k + 1)$ radicals, where k is a non-negative integer. Such systems are easy to identify as being of open-shell type by their odd number of C atoms. The present procedure (Coulson & Rushbrooke, 1940) divides all of the C atoms corresponding to the vertices of the schematic diagram into a set of starred and a set of un-starred atoms, such that no atom of one set is adjacent to an atom of the other. The rule is then that:

> If the numbers of atoms in the starred and un-starred sets are equal, then the compound is normal (closed-shell). If not, the compound is an open-shell n-radical and n is the difference between the two sets.

The use of this rule is demonstrated by Figures 26 and 27. Zethrene (Figure 26) is a normal compound since the numbers of starred and un-

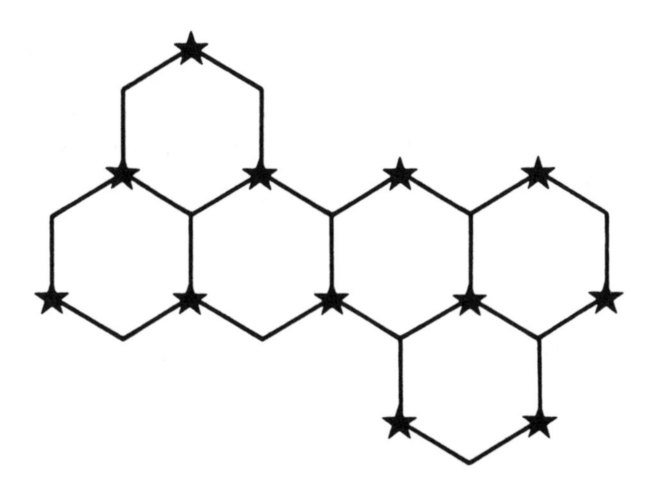

Figure 26. The use of the Coulson-Rushbrooke rule to determine whether a benzenoid hydrocarbon is of open- or closed-shell type. Here, zethrene is seen to be a normal (closed-shell) compound since the numbers of starred and un-starred C atoms are equal.

starred C atoms are equal. On the other hand, triangulene (Figure 27) is a di-radical because the difference in the numbers of starred and un-starred atoms is equal to 2. When the structure is very large, Balaban

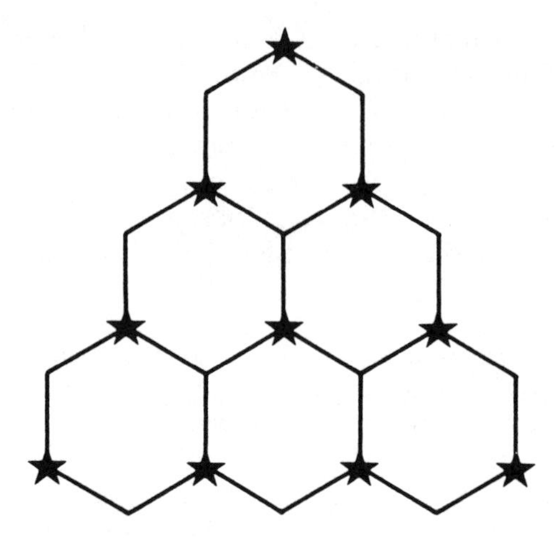

Figure 27. The use of the Coulson-Rushbrooke rule to determine whether a benzenoid hydrocarbon is of open- or closed-shell type. Here, triangulene is seen to be a di-radical because the difference in the numbers of starred and un-starred C atoms is equal to 2.

(1981) has suggested that it is easier first to triangulate the diagram (Figure 28) and then determine the difference in the numbers of triangles which have starred or un-starred atoms at their centers. In the case of ramified structures, such as the one shown, there is obviously much less danger of miscounting.

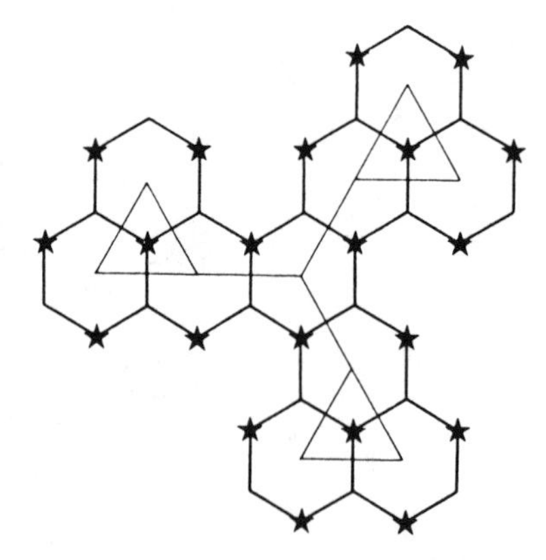

Figure 28. The use of the Balaban modification of the Coulson-Rushbrooke rule to determine whether a benzenoid hydrocarbon is of open- or closed-shell type. Here, a complicated perifusene is seen to be a tri-radical because the difference in the numbers of starred and un-starred triangles is equal to 3. (In this particular case, there are no un-starred triangles.)

CRAFT'S RULE

This is based upon the Clausius-Clapeyron equation for the correction of boiling points to atmospheric pressure. It states that:

$$\Delta T = c(273 + T_b)(760 - P)$$

where ΔT is the temperature difference which has to be added in order to give the boiling point at a pressure of 760mm, T_b is the boiling point observed at a pressure, P (mm), and the constant, c, is equal to 0.00012 for most liquids.

CRAM'S RULE

The difference in energy between diastereoisometric transition states, and the resultant predominance of one diastereoisometric product over the other, is most marked when the newly created asymmetric center is close to an asymmetric center which is already in the molecule. In the particular case where the two centers are adjacent to each other and where asymmetry at the new center is created by an addition reaction to a double bond, the present rule predicts which will be the predominant stereoisomer in the product. The rule (Cram & Abd Elhafez, 1952) is most easily understood by referring to a diagram (Figure 29), and states that:

When the asymmetric C atom is oriented so that the carbonyl function is flanked by the two smaller groups (M, S) attached to C, the reagent (R'X) preferentially approaches the carbonyl group from the side of the smallest group (S).

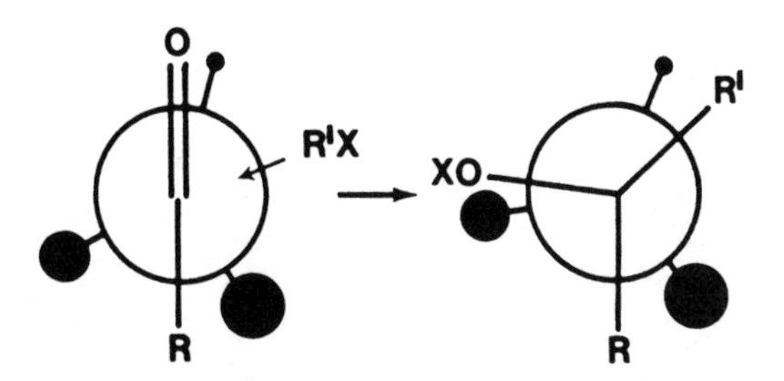

Figure 29. Cram's rule. When a C atom is oriented so that the carbonyl function is flanked by two smaller groups (M, S) attached to C, the reagent (R'X) preferentially approaches the carbonyl group from the side of the smallest group (S).

This rule applies only to reactions which are kinetically controlled. That is, it applies to the product which is formed in a rate-controlled process; not to the more stable product which is formed during subsequent equilibration.

CRUM-BROWN & GIBSON RULE

If a compound can be directly oxidized it is meta-directing when substituted into a benzene ring. If it cannot be oxidized, it is ortho-para directing.

For example, the NO_2 group is meta-directing because nitrous acid is directly oxidized to nitric acid. However, the OH group is ortho-para directing since water cannot be directly oxidized to give hydrogen peroxide.

CUBE-PER-ORDER INDEX RULE

There are two common features of all laboratories. These are a large amount of equipment and a limited amount of space. If, at some time, the reader should find himself in charge of storing such equipment, this storage assignment rationale (Kallina & Lynn, 1976) may be of use. It states that,

The item with the lowest ratio of space requirement to transaction demand should be placed in the nearest available location.

Here, the space requirement is the item's volume and the transaction demand is the number of requests for that item.

DALTON RULE

See Dühring's rule.

DARZENS' RULE

It was suggested (Darzens, 1897) that:

$$L_e/T_b = f(T_b/T_c)$$

where L_e is the latent heat of evaporation, T_b is the boiling point, T_c is the critical temperature, and f is an unspecified function; which might be linear, or even constant, in some cases.

DE FORCRAND'S RULE

It was suggested (de Forcrand, 1901) that:

$$(L_e + L_f)/T_1 = \text{constant}$$

where L_e is the latent heat of evaporation, L_f is the latent heat of fusion, and T_1 is the temperature in degrees Kelvin at which the vapor pressure is equal to 1atm.

DE HEEN'S RULE

According to this rule (Chwolson, 1923):

$$c_{pl} - c_{pg} = 4aL/3$$

where c_{pl} is the specific heat at constant pressure for a liquid, c_{pg} is the specific heat at constant pressure for a vapor, a is the coefficient of expansion, and L is the latent heat of evaporation.

DE MALLEMANN'S RULE

The additivity of optical rotations (see Biot's rule) is often used to determine the composition of a mixture. However, a complication arises due to the fact that the specific optical rotation depends upon the wavelength of light in a manner which is different for each component of the mixture, and for the mixture itself. The analysis of such data is simplified by the use of the present rule (De Mallemann, 1923), which states that:

When tangents are taken to dispersion curves, at points corresponding to the same wavelength, all of the tangents pass through a single point.

Here, the dispersion curves are plots of specific optical rotation versus the wavelength the light used, and the curves are usually those for two pure components of a mixture and for the mixture itself.

DIAGONAL RULE

Similarities in chemistry exist between the first-row elements and the elements which are found diagonally to their lower right in the periodic table.

This correlation, particularly between the properties of Li and Mg, Be and Al, B and Si, has been recognized for some 120 years. It is commonly attributed to systematic changes in the ionic potential. The latter is equal to the charge on the ion, divided by the ionic radius, and it is suggested that these parameters change in going down (increasing radius) and across (increasing charge) the periodic table in such a way as to maintain a constant value of the ionic potential. Unfortunately, the values of the ionic potentials (Table 6) do not in fact reflect this supposed constancy. See also the Shirley-Hall rule.

DINGLE'S RULE*

An increasingly important trend in semiconductor technology is the tailoring of semiconducting properties by the fabrication of layer structures. However, it is necessary to satisfy electronic constraints at the interfaces. The factors to be considered include transport across the heterointerface, the sub-band energy of quantum-well structures, and band discontinuity. It was deduced (Dingle, 1975), on the basis of absorption spectrum measurements, that:

> The conduction band discontinuity at an interface between two different semiconductors is equal to 85% of the difference in the band gaps of the adjacent semiconductors.

See also Anderson's rule, Bastard's rule, and the Furdyna-Kossut rule.

Table 6
Ionic Potential of Common Cations
(see the diagonal rule)

Ion	Coordination	Z/r
Al^{3+}	4	77
B^{3+}	4	273
Ba^{2+}	6	15
Be^{2+}	4	74
C^{4+}	4	267
Ca^{2+}	6	20
Cd^{2+}	4	21
Cs^{+}	6	6.0
Ga^{4+}	4	64
Ge^{4+}	6	55
Hg^{2+}	4	21
In^{3+}	4	48
K^{+}	4	7.3
La^{3+}	6	29
Li^{+}	4	17
Mg^{2+}	4	35
Na^{+}	4	10
Pb^{4+}	6	34
Rb^{+}	6	6.6
Sc^{3+}	6	40
Si^{4+}	4	154
Sn^{4+}	6	58
Sr^{2+}	6	17
Tl^{3+}	4	40
Y^{3+}	6	33
Zn^{2+}	4	33

DIPOLE RULE

See the Van Arkel rule.

DISPLACEMENT RULE

See Freudenberg's rule.

DISTANCE RULE

If the first member of a homologous series is laevorotatory and the third member is dextrorotatory, for instance, the rotation values of the first and the third member should be more laevo and dextro, respectively, than that of the second.

DO-YEN-CHEN RULE*

$$T_c = A(T_m + T_b)$$

where T_c is the critical temperature, T_m is the melting point, and T_b is the boiling point (appendix 1); all in degrees Kelvin.

The authors (Do et al., 1984) noticed that Prudhomme's rule (qv), where A is 1, and the Lorenz rule (qv), where A = 0.9, gave good predictions for non-metallic elements and compounds but did not apply to metallic elements. They found that better predictions were obtained by setting A equal to other values, such as 1.2 for groups IVB, VB, or VIB, 1.3 for group IB, 1.6 for group IA, and 1.9 for group IIB. See also Porlezza's rule.

DOHERTY'S RULE*

On the basis of correlations which were found between heterogeneous nucleation activity and the melting point ratio of binary metallic systems (Doherty, 1978), it was suggested that:

For a catalyst, A, to be suitable for the grain refinement of an alloy, B, there should be, (i) metallic bonding in A, (ii) a high ratio of the melting point of A to the melting point of B, (iii) a low-energy solid/liquid interface.

DONALD-DAVIES RULE*

Some essential requirements for glass formation in metallic alloys are rapid quench rates and a composition which is close to a deep eutectic. A rule for glass formation has been proposed (Donald & Davies, 1978) which is based upon an arbitrary definition of a deep eutectic. This states that metallic glass formation will occur as the result of quenching processes if:

dT ($= 1 - T_L/T_M$) is greater than 0.2

where T_L is the liquidus temperature of the alloy and T_M is a weighted average of the melting points of the constituents of the alloy.

The rule appears to be quite accurate when applied to binary alloys (Table 7) and, in principle, can be just as easily applied to ternary alloys if suitable phase diagrams are available.

Table 7
Donald-Davies Rule for Metallic Glass Formation
(Glass should occur if dT is greater than 0.2)

Alloy	dT	Glass
$Zr_{65}Be_{35}$	0.37	yes
$Hf_{56}Be_{44}$	0.32	yes
$Pd_{80}Be_{20}$	0.32	yes
$Ti_{62}Be_{38}$	0.27	[1]
$Sc_{55}Be_{45}$	0.20	no
$Be_{67}Si_{33}$	0.15	no
$Be_{90}B_{10}$	0.15	no
$Y_{61}Be_{39}$	0.14	no

[1] *Very high quench rate required*

DUFF'S RULE

π seconds is a nanocentury

One of a number of rules (Bentley, 1985) compiled for the convenience of computer programmers. For example, if a program takes 10^7 seconds to run, prepare to take a 4-month sabbatical.

DÜHRING RULE

This improvement of an earlier rule due to Dalton permits the calculation of the vapor pressure of an element, since:

The ratio of the absolute temperatures at which the vapor pressures of two similar substances are the same is a constant.

As an example, the boiling points of Zn and Cd are 1180 and 1038K, respectively, and the vapor pressure of Zn at 610K is $10^{-5}mm_{Hg}$. According to this rule, the temperature at which the vapor pressure of Cd is the same is given by, $T_{Cd}/610 = 1038/1180$. That is, the required temperature is 535K. The now-obsolete Dalton's rule suggested that the vapor pressures were the same when the two substances were an equal number of degrees above or below their respective boiling points.

DULONG-PETIT RULE

If no heat capacity measurement exists for a particular substance, the data can be estimated by using the rule that, at temperatures which are not too low:

$$C_v = 3R$$

where C_v is the specific heat at constant volume, and R is the gas constant. That is, C_v is about 6cal/molK.

In modern textbooks the above equation is referred to as being a "law," whereas older works call it a "rule". This is an example of the way in which almost any correlation may eventually be accepted as a scientific fact.

EBERHART'S RULE*

Having noted that surface tensions had been determined for almost every liquid metal (except the rare earths) whereas there was a marked scarcity of similar data on liquid metal oxides, Eberhart (1966) sought and found a correlation between the two types of data. This states that:

> For a metal oxide with the empirical formula, M_xO_y, the ratio, r, of the oxide's surface tension to the metal's surface tension is given by $r/(x + y) = \text{constant}$.

The value of the constant given by Eberhart was equal to 0.148. Most of the data which he presented followed the rule quite closely. The greatest discrepancy occurred in the case of B_2O_3.

ECKROTH'S RULE*

This (Eckroth, 1967) is a commonly used extension of the Von Bayer method as applied to bridged cyclic systems in organic chemistry. It says that:

> In order to facilitate the naming of bridged cyclic systems, first distort a planar projection of the system by changing the bond lengths and angles until no bonds overlap.

This rule has the drawback that, while it is easy to apply it to adamantane (Figure 30—see Reddy's rule), it requires considerable geometrical insight to be able to apply it to dodecahedrane (Figure 31). See also Trahanovsky's rule.

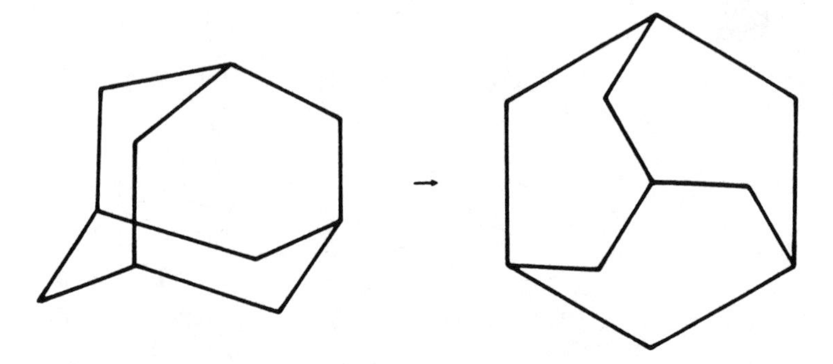

Figure 30. Illustrating Eckroth's rule for facilitating the naming of bridged cyclic systems by the distortion of bond lengths and angles, as applied to adamantane (tricyclic).

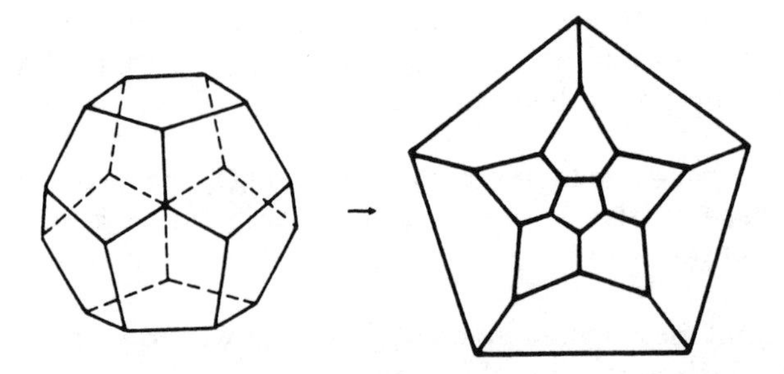

Figure 31. Illustrating Eckroth's rule for facilitating the naming of bridged cyclic systems by the distortion of bond lengths and angles, as applied to dodecahedrane (undecacyclic).

EGGERT'S RULE

It was proposed (Eggert, 1932) that the latent heat of evaporation varies in the following manner as a function of temperature:

$$L_e/(T_c - T) = \text{constant}$$

where L_e is the latent heat of evaporation, T is the temperature, and T_c is the critical temperature. The constant is supposed to be a universal one.

EICHLER-WILLE RULE

This is one of a number of correlation rules which relate the energy levels in separate atoms to those in the molecule. It is also one of the few which are not too complicated in themselves and do not link complicated integral functions of the atoms and molecules. Also, it is an improvement on the Barat-Lichten rule (qv). The present rule (Eichler & Wille, 1974) states that:

$$n_{u,r} = n_{i,a}$$

where $n_{u,r}$ is the number of zeroes of the radial wave function for the united atom and $n_{i,a}$ is the number of zeroes of the angular wave function for the isolated atom.

The rule is empirical, and its use in the construction of molecular orbital correlation diagrams calls for additional assumptions since the rule itself is satisfied by an infinite set of integrals. See also the Kereselidze-Kikiani rule.

EINSTEIN RULE

Workers in the field of photochemistry are concerned with the chemical changes which can be brought about by radiation with a wavelength of between 100 and 1000nm. This corresponds to photon quantum energies ranging from 23 to 230kcal/mol; which are also of the same order as the dissociation energies of the chemical bonds. Although the absorption of one quantum of radiation does not always produce one unit of the product, it is often assumed to a first approximation that:

One quantum of absorbed radiation activates only one molecule.

The compiler has been unable to identify the original statement of this principle, but a typical reference to it can be found in Reid (1968).

ELECTRON AFFINITY RULE

See Anderson's rule.

ELECTRON RULES

These are very general rules which have gradually been developed for the prediction or explanation of the structures of intermetallic compounds. They state that:

1. If the valence electron concentration and composition of two phases are the same, they often have the same, or very similar, structures.
2. If the valence electron concentration changes rapidly with composition, small ranges of homogeneity of the intermetallic phases are found.
3. If the valence electron concentration does not change rapidly with composition, wide ranges of homogeneity are found.

4. To a first approximation, valence electron concentrations of between 0 and 3 correspond to metallic bonding, concentrations of between 3 and 5 correspond to ionic and covalent bonding, and concentrations of between 5 and 8 correspond to molecular bonding.

These rules are the basis for some of those of Hume-Rothery (qv), and for the rules of Goryunova (qv) and Zintl (qv). They also largely overlap those of Engel (qv). Sets of such rules have also been formulated for special classes of compound. One such class is that of structures of the tetrahedral adamantine type. These structures can be of normal or defect type and have a close-packed completely occupied anion partial structure, while vacancies can occur on the cation sites. For this type of structure, the rules are that:

5. The number of vacancies in a defect adamantine structure is determined by the valence electron concentration.
6. The adamantine structure can occur only if both the general tetrahedral structure equation and the equation for normal valence compounds are satisfied.
7. The adamantine structure can occur only when the valence electron concentration is between 4 and 4.8.

ENGEL'S RULES

These (Brewer, 1932) are more specific rules than, and overlap, the electronic rules (qv). They state that:

1. The stability or binding energy of a solid depends upon the average number of unpaired electrons per atom which is available for bonding.
2. The contribution of the d-electrons to bonding increases with increasing atomic number of the metal in the periodic table, whereas the contribution of the sp-electrons decreases.
3. The crystal structure depends upon the number of s- and p-electrons present, giving body-centered cubic structures at less than 1.5 sp-electrons per atom, hexagonal closed-packed structures at between 1.7 and 2.1 sp-electrons per atom, and face-centered cubic structures at between 2.5 and more than 3 sp-electrons per atom.
4. The d-electrons do not directly determine the crystal structure.
5. The number of s-, p-, d-, and f-electrons in metal phases is close to that for gaseous atoms. If neighboring atoms have unpaired d-electrons available for bonding, their number of unpaired d-electrons is higher.
6. Carbon and nitrogen, in combination with transition metals of groups IV and V of the periodic table, convert the electronic structure to that of the more stable group VI elements.

The basic concepts above can be extended in many ways. For instance, they have been used (Fuchs & Ficalora, 1985) in order to deduce that the ultimate tensile strength of Ni-based superalloys is given by:

$$U = -618d + 4266$$

where U is the ultimate tensile strength in MPa and d is the d-electron concentration.

ENTHALPY OF MIXING RULE

This rule (Farlow et al., 1985) states that, in the ion-beam mixing of metals and insulators,

Metals mix with insulators if the reaction enthalpy is negative, and they do not mix if the reaction enthalpy is positive

EÖTVÖS RULE

According to this rule (Von Eötvös, 1886):

The rate of change of the molar surface energy as a function of temperature is the same for all liquids, and is independent of the temperature.

The proportionality constant is known as the Eötvös-Ramsay-Shield constant. Unfortunately, this rule is not as generally true as was originally thought. Some exceptions are hydroxyl-containing liquids and very low boiling point substances.

ERLENMEYER'S RULE

This states that:

The $-C(OH)_2-$ grouping is unstable, and gem-dihydroxy compounds do not usually exist.

Thus, the enolic isomerides of the simple aldehydes, ketones, and acids are so unstable that only the more stable ketonic forms can be isolated, although the existence of the former is deduced from the ability to prepare derivatives such as vinyl chloride.

FAJANS' RULES

It was suggested (Fajans, 1931) that the gradual transition from ionic to covalent bonding in a related series of compounds could be explained on the basis of the mutual deformation or polarization of the ions. According to this theory, when negative ions approach positive ions during compound formation the electron systems of the negative ions are shifted towards the positive ions by varying amounts. These shifts partially decrease the extent of charge separation in the original ions and lead to some covalent bonding. These considerations lead to the following rules for predicting bond type:

1. An increased charge on the anion leads to greater polarization and promotes covalent bonding.
2. A decreased size of the cation or an increased size of the anion produces greater deformation of the anion.
3. Non-rare gas ions are more efficient polarizers and the bonds which they form are more covalent.

In an interesting extension of these rules, Stone (1984) has introduced the concept of Fajans' "arrow." The latter is a line, on the periodic table, which joins the two components of a binary compound. It is suggested that the angle of the arrow can be used to predict the transition temperature for superconductivity.

FAJANS' PRECIPITATION RULE

A radioactive element will be co-precipitated with another substance if the precipitation conditions are such that the element would form a sparingly soluble compound if it were present in weighable quantities.

As it stands, there are many exceptions to the rule. However, in modified form (see Hahn-Fajans rule) it is more generally true. See also Paneth's adsorption rule.

FERMI RULE

In his selection rules for beta decay, Fermi suggested that the total angular momentum of a nucleus was unchanged by the emission of a beta particle, and that there was no change in parity. That is, the emitted beta particle and neutrino had opposite spins, and there was no change in right/left symmetry. The rule must be applied with caution because it was later found that the beta particle and the neutrino could have parallel spins. The Gamow-Teller (qv) rules were introduced in order to replace it.

FILONENKO'S RULE*

This rule (Filonenko, 1970) is useful in predicting the likely microstructure of a eutectic alloy:

> If the ratio of the entropies of fusion of the two component phases is less than 1.5, the microstructure will be regular. If the ratio is greater than 1.5, the microstructure will be irregular.

The exact nature of the microstructure is best revealed by the use of directional solidification.

FINAL-STATE RULE

Many-body effects make it difficult to interpret X-ray absorption data and emission spectra on the basis of one-particle band structures. However, it is known (Von Barth & Grossman, 1979) that:

> Observed spectra represent the final state of the system, and the $L_{2,3}$ emission spectra of simple metals can be estimated by calculating the dipole matrix elements on the assumption that there is no hole in the 2p level.

FINE-BROWN-MARCUS RULES*

Empirical relationships were found (Fine et al., 1984) between the elastic constants and melting points of various materials (Table 8). These led to the rule that:

$C_{11} = 0.17T_m - 93.6$

for metals and intermetallic compounds, where T_m is the melting point in degrees Kelvin and the elastic constant is measured in GPa.

The bulk modulus is given by:

$B = 0.11T_m - 65.3$

for the same materials.

Table 8
Elastic Constants and Melting Points of Various Materials

Material	C_{11}(GPa)	B (GPa)	T_m(K)
Al_2Ca	97	47	1352
Al_2Gd	161	78	1798
Al_2La	144	69	1697
Al_2U	170	83	1863
Al_2Y	171	80	1758
AuCd	83	85	900
AuZn	42	56	998
Co_2Hf	256	167	1843
Co_2Zr	233	153	1813
Cu_2Mg	123	88	1092
CuZn	129	116	1175
LiIn	56	47	898
MgAg	85	66	1093
MgCuZn	134	58	1098
Mg_2Si	121	55	1375
Mg_2Sn	82	41	1051
NiAl	212	166	1911
Ni_3Al	224	174	1668
NiTi	162	140	1513
Sn_3Ce	81	55	1435
V_3Ge	297	168	2193
YZn	94	62	1378
B_6La	453	90	2803
TaC	505	217	4073
TiC	515	242	3523
UC	320	163	2663
ZrC	472	168	3450

However, this rule is less well obeyed because the correlation between the C_{12} elastic constant and temperature is less good. In the case of materials with the diamond cubic or GaAs structure, the rule is:

$$C_{11} = 0.12T_m - 50.2$$

Rules of this type are not as accurate in predicting the C_{11} value of III-V semiconductors as they are in predicting this value for intermetallic compounds.

FINNIE-WOLAK-KABIL RULE*

It has been found (Finnie et al., 1967) that:

The solid particle erosion rate of annealed face-centered cubic metals is inversely related to their hardness.

See also the rules of Ascarelli, Brauer-Kriegel, Hutchings, Khruschov, Smeltzer, and Vijh.

FINUCAN'S RULE*

This rule (Finucan, 1973) is very useful for solving problems, such as those involving Markov processes, which lead to sets of linear iterations having a particular form. That is, to equations of the form:

$$p_{n+1} = ap_n + bq_n + cr_n$$
$$q_{n+1} = dp_n + eq_n + fr_n$$
$$r_{n+1} = gp_n + hq_n + ir_n$$

with the conditions that $a + d + g = 1$, $b + e + h = 1$, and $c + f + i = 1$.

It is often desired to find the limiting values of p and q to which these iterations lead, and it is here that the present rule is helpful. It can be stated in the form:

Write the transition matrix of the equations, reduce each diagonal element by unity, and take the components in the ratio of the principal minors.

The use of this rule can be quickly appreciated by treating a concrete example. If a = 0.2, b = d = h = 0.3, c = e = 0.4, f = g = 0.5, and i = 0.1 in the above equations, then the transition matrix (first part of rule) is:

$$\begin{matrix} 0.2 & 0.3 & 0.4 \\ 0.3 & 0.4 & 0.5 \\ 0.5 & 0.3 & 0.1 \end{matrix}$$

Reducing the diagonal elements by unity (second part of rule) gives:

$$\begin{matrix} -0.8 & 0.3 & 0.4 \\ 0.3 & -0.6 & 0.5 \\ 0.5 & 0.3 & -0.9 \end{matrix}$$

and the principal minors corresponding to (1,1), (2,2), and (3,3) are:

$$(-0.6 \times -0.9) - (0.3 \times 0.5) = 0.39$$
$$(-0.8 \times -0.9) - (0.5 \times 0.4) = 0.52$$
$$(-0.8 \times -0.6) - (0.3 \times 0.3) = 0.39$$

Taking the limiting p, q, and r values in the ratio of the minors (third part of rule) and recalling that they should add up to unity, gives:

$$p = 0.3$$
$$q = 0.4$$
$$r = 0.3$$

This rule always gives the correct answer, and is applicable to transition matrices of any order.

FLEMING'S RULE

A well-known rule which permits one to quickly predict the relative directions of movement, electron flow, and magnetic flux in an electrical device. The index finger, second finger, and thumb are placed at 90° to each other and are taken to represent the magnetic flux (B) direction, electromotive force (E), and mechanical torque (T), respectively. If the right hand is used, the conditions are those for a generator, whereas the left hand gives the relationship between those quantities for a motor.

FLYNN'S RULE*

An increasing field of interest is the preparation of superlattices by using various epitaxial methods. The present rule (Flynn, 1988) states that,

> In order to obtain high-quality epitaxial interfaces, the growth temperature should be equal to about $3T_m/8$, and the components should have similar melting points.

Here, T_m is the melting point on the absolute temperature scale.

FRANKLAND RULE

It was a very early suggestion (Frankland, 1899) that:

> In aliphatic carbon chains, the carbon atoms form spirals, and a turn of the spiral is completed with 5 carbon atoms. In homologous series, compounds with 5 (or a multiple of 5) chain units should exhibit anomalous physical constants; especially optical rotation.

Later work confirmed the existence of anomalous rotatory powers. Thus, in the homologous series of methyl-n-alkyl carbinols, the optical rotation of the propyl compound (with n = 3) is abnormally high. In benzene, or in alcoholic solutions, the amyl (n = 5), octyl (n = 8), and decyl (n = 10) compounds exhibit anomalies.

FREUDENBERG'S RULE

The study of the optical rotatory powers of organic compounds is a powerful method for deducing their structures. In particular, simple and reasonably accurate rules have been developed which permit a "relative" configuration to be assigned. One of these is the Van't Hoff rule of optical superposition (qv). Another rule is the present one, which is also known as the displacement rule or the shift rule (Freudenberg et al., 1930). This states that:

> When two similarly constituted disymmetric compounds are changed chemically in the same way, and the change produces a considerable shift in optical rotation in the same direction, then the two compounds probably have the same configuration.

The use of this rule is illustrated by Table 9. Here, the changes in the optical rotatory powers of derivatives of mandelic acid (known structure) are compared with the corresponding changes for atrolactic acid ("unknown" structure). The fact that most of the changes are in the same direction strongly suggests that the two structures are the same.

Table 9
The Use of Freudenberg's Rule for Determining the Molecular Structure of an Organic Compound

Compound	Derivative	Rotation(°)	Shift Direction
mandelic acid	...	– 240	...
atrolactic acid	...	– 86	...
mandelic acid	ethyl ester	– 210	+
atrolactic acid	ethyl ester	– 59	+
mandelic acid	amide	– 137	+
atrolactic acid	amide	9	+
mandelic acid	benzoyl-ethyl ester	– 404	–
atrolactic acid	benzoyl-ethyl ester	– 67	+
mandelic acid	acetyl-ethyl ester	– 280	–
atrolactic acid	acetyl-ethyl ester	– 40	+
mandelic acid	amide acetone derivative	– 181	+
atrolactic acid	amide acetone derivative	103	+
mandelic acid	methyl ether methyl ester	– 197	+
atrolactic acid	methyl ether methyl ester	110	+
mandelic acid	methyl ether dimethyl ester	290	+
atrolactic acid	methyl ether dimethyl ester	395	+

FRIEDEL'S SCREENING RULE

When considering the electronic structures of point defects, it is common (Koch & Koenig, 1986) to use the approximation that:

> Since the self-consistent potential at the defect has to screen a charge which is sufficient to ensure global neutrality of the crystal, the total displaced charge up to the Fermi level must be equal to the difference in the atomic numbers of the defect and the replaced atom at the origin.

FRIEDERICH'S RULES

The problem of predicting the electrical properties of a solid on the basis of its chemical composition and crystal structure is an old one. On the basis of the experimental data then available to him, Friederich (1925) came to the conclusion that:

1. In compounds, the non-saturated valence of the metal atom gives rise to electronic conduction; corresponding compounds with saturated bonds exhibit very poor electrical conduction.
2. Free valences (free-valence electrons) lead to electrical conduction.
3. Only compounds which do not exhibit electronic conduction can have an ionic structure.
4. Compounds with molecular structures do not have free conduction electrons.

FRIEDERICH-MAYER RULES

The general behavior of oxide semiconductors is summarized by these empirical rules (Friederich, 1925; Mayer, 1928):

1. The electronic conductivity of oxides derived from the highest valency state is increased when the partial pressure of oxygen in contact with the oxide is decreased.
2. The electronic conductivity of oxides derived from the highest valency state is decreased when the partial pressure of oxygen in contact with the oxide is increased.
3. The electronic conductivity of oxides based upon the lower of two possible valency states is increased when the partial pressure of oxygen in contact with the oxide is increased.
4. The electronic conductivity of oxides based upon the lower of two possible valency states is decreased when the partial pressure of oxygen in contact with the oxide is decreased.

The rule can be applied to industrial processes, such as sintering, if the mechanism depends upon the diffusion of negative or positive holes. In this case, an oxide which is based upon the highest valency state of the metal should be more easily sintered if the oxygen partial pressure of the surrounding atmosphere is decreased.

FRIES' RULE

The most stable arrangement of the bonds of a polynuclear compound is that in which the maximum number of rings has the benzenoid arrangement of 3 double bonds.

This means that, in the stated case, there will be three double bonds in each ring.

FURDYNA-KOSSUT RULE*

An increasingly important trend in semiconductor technology is the tailoring of semiconducting properties by the fabrication of layer structures. However, in order to avoid the generation of defects such as dislocations at the interfaces it is desirable to match the two lattices as closely as possible. A simple rule (Furdyna & Kossut, 1986) was developed for the case of tetrahedrally bonded semiconductors:

> If two compound semiconductors of the form, A(a)B(b) and A(c)B(d), have a common interface, then good lattice matching will occur if a = c and b = d or if a = d and b = c.

Here, the letters in parentheses indicate rows of the periodic table, and the nature of A and B is governed by rules such as that of Zintl (qv). The present rule is based upon the additivity of the tetrahedral radii of the component atoms, and upon the fact that these radii hardly vary in going along the row. The use of the rule is best seen by means of a concrete example. In the case of a CdSe/InAs heterojunction, the atomic radii are Cd (0.1405nm), Se (0.1225nm), In (0.1405nm), and As (0.1225nm). Therefore, the theoretical bond lengths are CdSe = InAs = 0.2630 (= 0.1405 + 0.1225). Zero mis-match is predicted and, indeed, the experimentally measured value of 0.08% is one of the lowest found in such heterostructures. It is important to note that the arrangement of elements in the periodic table is not followed exactly. Thus, because Al has almost the same radius as the element in the row below (Ga) it is moved to the lower row for the purposes of this method. The rule seems to be capable of wide application. See also the Connection rules.

FÜRST-PLATTNER RULE

> In the case of epoxides with a rigid structure, an entering substituent and the hydroxyl group which is formed occupy the axial position.

See also the Conformational rules.

g-PERMANENCE RULE

This rule (Pauli, 1923) is applicable to the terms of a multiplet of an atom and states that:

The sum of the g-factors is the same for strong and weak magnetic fields, for a given value of M.

Here, M is one of the magnetic quantum numbers corresponding to a particular value of J. For instance, when $J = 1/2$, the M values are $1/2$ and $-1/2$. The doublet terms, $^2r_{-3/2}$ and $^2r_{-1/2}$, have $1/2$ and $-1/2$ as common M values. The rule implies that for $M = 1/2$, say, the sum of the g-factors for the two terms is the same in strong and weak magnetic fields. In the present case, the sum is equal to 2.

GAMMA PERMANENCE RULE

For a particular spectral multiplet (same L and S) in L-S coupling, or same j_1 and j_2 in j-j coupling, and for the same value of the magnetic quantum number, M, the sum of all of the gamma factors is the same for both strong and weak magnetic fields.

For example, the doublet terms, $^2P_{3/2}$ and $^2P_{1/2}$, have M values of $1/2$ and $-1/2$ in common. The rule implies that for $M = 1/2$, say, the sum of the g-factors for the two terms is the same in strong and weak magnetic fields. In the present case, the sum is equal to $-1/2$. This rule is very similar to the previous one, but they are given separately because it is usual to attribute the g-permanence rule to Pauli and the gamma permanence rule to Landé.

GAMMA SUM RULE

This is an extension (Goudsmit, 1928) of the gamma permanence rule and states that:

For given quantum numbers of the electrons, the sum of the gamma values belonging to a given total moment, j, is independent of the type of coupling of the electrons.

This is true for both weak and strong magnetic fields.

GAMOW-TELLER RULE

In these selection rules for beta decay, it is suggested that:

The change in total angular momentum of a nucleus is equal to $\pm h/2\pi$ or zero, with no change in parity, and the initial or final value of the total angular momentum cannot be zero.

Like the Fermi rules (qv), the present ones must be used with caution because it is now known that parity is not always conserved in "weak" interactions such as beta decay.

GEOMETRIC COMBINING RULE

Combining rules are often used (Stwalley, 1971) to deduce the interaction potential, P_{AB}, of asymmetric pairs of atoms or molecules when those of the symmetric pairs (P_{AA}, P_{BB}) are known. One of the most commonly used is the rule that:

The asymmetric pair interaction potential between two atoms or ions is given by:

$$P_{AB}(R) = [P_{AA}(R)P_{BB}(R)]^{1/2}$$

where $P_{AA}(R)$ is the A-A interaction potential at separation, R, etc.

However, this rule has a disadvantage when the difference in the sizes of A and B is large. This is because the interactions in all three systems are assumed to occur at equal values of R. In order to avoid this problem, Smith (qv) and Lee & Kim (qv) proposed combining rules, for the repulsive interaction, which take account of this size difference. When a geometric mean value is used to estimate the effect of molecular size on surface tension, it is called the Berthelot rule. See also the arithmetic combining rule.

GEODESIC LENS RULE*

This rule, which was developed (Sottini et al., 1979) for screening the geodesic lenses used in integrated optical processors, permits an accurate evaluation of the resultant change in focal length due to a depth change arising from fabrication errors. It states that:

$$df/f = Q(dz/z)$$

where df/f is the relative change in the focal length of the lens, dz/z is the relative error in the depth of the lens, and Q is a constant which is approximately equal to 1.9.

GIAMBIAGI RULES*

Two simple rules were presented (Giambiagi & Giambiagi, 1979) for the bond order properties of conjugated organic molecules having two-fold symmetry. They state that:

1. Doubly-charged negative forms of the molecule have electronic pi-charges of 2 for atoms on the axis, and unit pi-charges for the other atoms. The mobile bond orders are 1, if they relate symmetrically equivalent centers, and are zero otherwise.
2. Doubly-charged positive forms of the molecule have electronic pi-charges of 0 for atoms on the axis, and unit pi-charges for the other atoms. The mobile bond orders are -1, if they relate symmetrically equivalent centers, and are zero otherwise.
3. In systems which possess a two-fold symmetry axis which does not pass through any pi-center, the summation of the bond orders crossing the symmetry axis is equal to half of the difference between the number of electrons occupying symmetric and antisymmetric levels.

The last rule holds not only for the ground state, but also for the excited states.

GIBBS' RULE

The pressure of a mixture of gases is equal to the sum of the partial pressures only if the chemical potential of the gas is unchanged by being incorporated into the mixture.

This is a more rigid version of the better-known Dalton's law of partial pressures.

GILBERT'S RULES*

When considering the kinetic behavior of chemical reactions, it is frequently necessary to derive the rate law for a proposed mechanism in terms of the individual kinetic constants. A simplification of the rate law is often obtained when the steady-state approximation can be used. Even when this approximation can be applied, the derivation of the rate expression for reaction mechanisms involving more than one intermediate product is tedious. However, Gilbert (1977) presented a simple method for deriving the steady-state rate equations for first-order and pseudo-first order reactions. The method permits the derivation of the observed first-order rate constant in terms of the individual kinetic constants, merely by inspection. The rules state that:

1. The numerator for the overall forward reaction is the product of the n individual rate constants.
2. The denominator for the overall forward reaction is the sum of n terms; each of which is the product of (n - 1) rate constants.
3. Write the mechanism down. Place the left thumb over one of the steps. Write the product of the forward rate constants for steps to the right of the thumb, multiplied by the reverse rate constants for all of the steps to the left of the thumb. Repeat this procedure for the remaining steps. The denominator is then the sum of all of the terms generated in this way.

Note that this is quite literally a "rule of thumb;" to such an extent that the original paper sported printed thumb-prints in the figures. The expression for the overall reverse rate constant is simply the product of all of the reverse rate constants, divided by the same denominator as that for the overall forward rate constant. The ratio of the overall forward rate constant to the overall reverse rate constant is the equilibrium constant for the overall reaction. The sum of the overall forward and reverse rate constants is the observed rate constant for the approach to equilibrium from either direction. The use of the rules is seen more clearly by considering a typical mechanism (Figure 32). An extension of the rules can be made when just one step is rate determining; giving a fourth rule:

4. If one step is entirely rate-determining, the rate constant for the reaction in the forward direction is the product of all of the equilibrium constants for the steps preceding the rate-determining step, multiplied by the rate constant for the rate-determining step.

The rules can also be applied to cyclic reactions by breaking them down into a suitably pseudo-linear form and then applying the above methods.

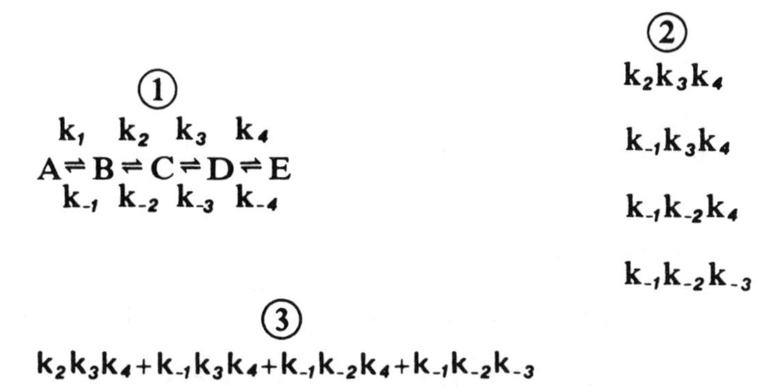

Figure 32. The use of Gilbert's rules to evaluate the reaction rate constants for complicated reaction mechanisms. Here (1), the k_i are forward rate constants and the k_{-i} are reverse rate constants. The numerator of the overall forward rate constant is the product of the individual rate constants. By covering up (with the thumb?) each reaction in turn, one obtains four terms (2) using rule 3. Adding these terms (3) gives the denominator. Dividing the numerator by the denominator gives the overall forward rate constant. The reverse overall rate constant can be found in the same way. The sum of the overall rate constants is the observed rate constant, and dividing the forward rate by the reverse rate gives the equilibrium constant.

GOLDSCHMIDT'S RULE

The coordination number of a metal tends to decrease with increasing temperature.

For example: Ti, Zr, Hf, Th, and Tl have 12-fold coordination at low temperatures, but 8-fold coordination at high temperatures. On the other hand, there are many cases in which the reverse behavior is exhibited; such as Sn, Cr, Fe, Mn, and Pu. The rule (Laves, 1967) is therefore of doubtful utility.

GOODENOUGH-KANAMORI RULES

1. When two ions have the lobes of their magnetic orbitals pointing towards each other in such a way that the orbitals would have a reasonably large overlap integral, then the exchange is antiferromagnetic.
2. When the orbitals are arranged in such a way that they are expected to be in contact but to have no overlap integral (especially when the overlap is zero by symmetry), then the expected behavior is ferromagnetic.

In the second case, the ferromagnetic interaction is not usually as strong as the antiferromagnetic one. These rules were given by Anderson (1963).

GOODSTEIN'S RULES*

In order to rationalize the use of electronegativity and oxidation/reduction principles, it was suggested (Goodstein, 1970) that one should use the rules that:

1. The oxidation number of a given atom equals the sum of the electrons bonding the given atom to atoms of higher electronegativity, less the sum of the electrons bonding the given atom to atoms of lower electronegativity.
2. In an oxidation/reduction reaction, a change in the relative atomic electronegativities takes place between a given atom and the atoms to which it is bonded. At the same time, and elsewhere among the reactants, there is an opposite change in the relative electronegativities. The difference in the relative electronegativities can be considered to be one of the factors which is causing the change to occur.

GORECKI'S RULES*

On the basis of an analysis of elastic property data for metallic elements (appendix 2), it was deduced (Gorecki, 1980) that:

The elastic moduli of a polycrystalline metal obey the relationships: G/E = constant, G/B = constant, and E/B = constant, where G is the shear modulus, B is the bulk modulus, and E is the Young's modulus.

The value of the constant depends upon the ratio in question and upon the structure of the metal (Table 10). It is also suggested (Gorecki, 1974) that vacancy parameters are related to the elastic constants and that:

$$E_f = U/3$$

$$E_f/T_m = \text{constant}$$

where E_f is the vacancy formation energy, T_m is the melting point, and U is the bonding energy per atom. In the second rule, the constant is equal to 0.833meV/K.

Following earlier work by others, correlations were found (Gorecki, 1979) between various surface energies:

E_{sv}/E_{lv} = constant

where E_{sv} is the solid/vapor surface energy at the melting point, E_{lv} is the liquid/vapor surface energy at the melting point, and the constant is equal to 1.18 for A1- or A3-structured metals and 1.20 for body-centered cubic metals.

Likewise:

$E_{sl}/E_{lv} = 0.142$

for all metals, where E_{sl} is the solid/liquid interfacial energy.

Table 10
Constants for Gorecki's Rule

Structure	Ratio	Constant
bcc	G/E	0.357
bcc	G/B	0.373
bcc	E/B	1.041
fcc	G/E	0.385
fcc	G/B	0.379
fcc	E/B	0.944
hcp	G/E	0.389
hcp	G/B	0.527
hcp	E/B	1.314

GORMAN'S RULES*

The use of Russell-Saunders symbols is common in inorganic chemistry and there are simple rules for writing the ground state symbols of free atoms and ions. It first has to be recalled that the net angular momentum of electrons in filled levels and sub-levels is zero and that attention thus only has to be paid to partially filled levels. The following rules can then be used to determine the ground state of a free atom or ion:

1. Write the electronic configuration of incomplete sub-levels.
2. List horizontally the m_l values[1] for the relevant incomplete s, p, d, and f levels; beginning at the left with the highest value[2].
3. Fill the orbitals identified in rule 2 by using the available electrons indicated by rule 1. Single electrons are added to each orbital before any pairing is done; beginning at the left. When pairing is necessary, this also starts at the left[3].
4. The m_l values of unpaired electrons are added algebraically to get a resultant value. If M_L is negative, the minus sign is dropped. The numerical value of M_L is assigned a letter according to the scheme: 0 = S, 1 = P, 2 = D, etc. However, J is not used.
5. The spectroscopist's spin multiplicity is indicated by a number which is one more than the number of unpaired electrons. This is used as a pre-superscript.

In order to clarify the use of these rules, it is helpful to see how they are applied to typical examples. In the case of the P atom, one first obtains $3p^3$ (rule 1), the m_l are 1, 0, -1 in that order (rule 2) and are all spin-up (rule 3). The M_L value is zero; an S state (rule 4). Finally, the multiplicity is $3 + 1$ (rule 5), giving a quartet, ^4S. In the case of Ti(III), one first obtains $3p^1$ (rule 1), the m_l are 2, 1, 0, -1, -2 in that order (rule 2) and only $m_l = 2$ is occupied (rule 3). The M_L value is 2; a D state (rule 4). Finally, the multiplicity is $1 + 1$ (rule 5), giving a doublet, ^2D. It should be remembered that these rules are not absolute, and that there may be exceptions.

GORYUNOVA'S RULE *

This is a rule which was deduced (Goryunova, 1965) from an empirical map of the Mooser-Pearson type, and is used to predict the occurrence of intermetallic compounds having the adamantine structure. It states that:

Adamantine structures will occur when $E_c - 0.1E_a$ is greater than 6

where E_c is the specific electron affinity of the cation, and E_a is the specific electron affinity of the anion.

[1] *Here, m_l is the quantum number which corresponds to the component of orbital angular momentum along a reference direction for a single electron.*

[2] *That is, $s = 0$, $p = 1$, $d = 2$, $f = 3$*

[3] *This is simply an application of Hund's rule (qv) for maximizing the total orbital angular momentum and the spin multiplicity; thus minimizing repulsion for the ion or atom as a whole.*

GROSHAN'S RULE

According to this very old rule (Groshan, 1847):

The absolute boiling points of organic liquids at two fixed pressures have a constant ratio of 0.78.

This conclusion is obviously subsumed by the various modern expressions which relate vapor pressure to temperature.

GRUNBERG-NISSAN RULE*

It is suggested (Grunberg & Nissan, 1948) that, for normal paraffin hydrocarbons,

$$T_c^3/M = \text{constant}$$

where T_c is the critical temperature and M is the molecular weight.

GULDBERG'S RULE

$$T_b/T_c = 2/3$$

where T_b is the boiling point and T_c is the critical temperature.

This rule (Guldberg, 1890) was independently proposed by Guye, and can be derived from Trouton's rule (qv). It is sometimes also stated that $T_m/T_b = 2/3$, where T_m is the melting point, but this seems to be less well obeyed. Both forms are considerably less consistent than the related Prudhomme's rule (qv). Such "two-third rules (qv)" are ubiquitous and hint at a common basis.

GUNNING'S RULE*

This states that:

Scientific papers should have 45 words per sentence and 25 of these should have more than 2 syllables.

Several systems have been developed in order to judge the readability of a text, and the "Fog Index" is one of the most interesting. Its calculation requires three steps:

1. Record the average sentence length in a sample text of some 100 words and count the number of words per sentence.
2. Count the number of words having three or more syllables. (In order to avoid a distorted result, terms such as "nuclear magnetic resonance" should be counted as a monosyllabic "NMR.")
3. Add the latter two numbers and multiply by 0.4.

The result is the Fog Index, and is also the estimated reading age required to assimilate the text. Gunning used a nomogram in order to define further the meaning of the index (Figure 33). However, this hardly constitutes a handy rule of thumb. Instead, one can use it to deduce various ready-made rules, such as the one above. In the days before commercial word-processing programs were available, the present compiler wrote such a program for an early desk-top computer. One of its features was a sub-program which returned the Fog Index of a given text. Such programs are now a common optional feature of dedicated commercial word-processors. Programmers who still wish to write this type of program should note that it is easier to assume that the average syllable is 2.7 letters long than it is to attempt to divide words into syllables on a linguistic basis.

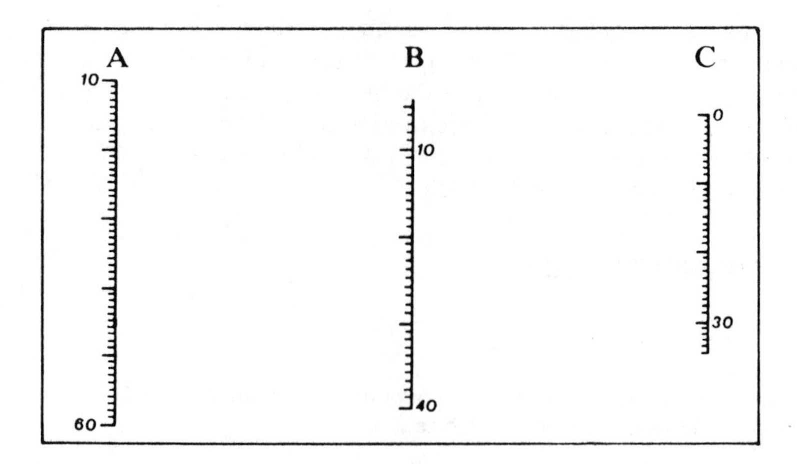

Figure 33. A nomogram used to gauge literary style in Gunning's rule. Here scale A is the average sentence length, scale B is the Fog Index (reading age), and scale C is the percentage number of words which have 3 or more syllables.

GUST-PREDEL RULE*

By analyzing self-diffusion data for high-angle grain boundaries in various metals, using a reduced-temperature Arrhenius plot, it was shown (Gust et al., 1985) that the results could be fitted by a single curve for each common class of metal structure. The curve was of the form:

$$dD = A \exp[-BT_m/T]$$

where dD is the product of grain boundary diffusivity and grain boundary thickness, T is the temperature, T_m is the melting point, and A and B are constants.

For face-centered cubic metals, A is equal to $9.7 \times 10^{-15} m^3/s$ and B is 9.07. For body-centered cubic metals, A is equal to $9.2 \times 10^{-14} m^3/s$ and B is 10.4. For hexagonal close-packed metals, A is equal to $1.5 \times 10^{-14} m^3/s$ and B is 10.3

GUTMANN'S RULES

A number of valuable rules concerning bond length and charge density variations in molecular donor-acceptor interactions have been formulated (Gutmann, 1978). These state that:

1. The smaller the intermolecular distance between the donor and the acceptor atoms, resulting from donor-acceptor interaction, the greater is the induced lengthening of the adjacent bonds in both the donor and the acceptor species.
2. A given sigma-bond within the donor or acceptor species is lengthened when, as a result of donor-acceptor interaction, the induced electron shift occurs from an atom carrying a positive fractional charge to one carrying a negative fractional charge. The bond is shortened when the electron shift is induced in the opposite direction.
3. As the coordination number of an atom increases, so do the lengths of all of the bonds which originate from the coordination center (Table 11).
4. Although donor-acceptor interaction will result in a net transfer of electron density from the donor species to the acceptor species it will, in the case of polyatomic species, lead to a net increase or pile-up of electron density at the donor atom of the donor species and to a net decrease or spill-over of electron density at the acceptor atom of the acceptor species.

In the case of the second rule, it is difficult to deduce the fractional charges on a given atom. It is suggested that electronegativity values should be used in order to estimate the relative signs of the charges.

Table 11
Demonstrating Gutmann's Third Rule.
(As the coordination number of an atom increases, so do the lengths of all of the bonds which originate from the coordination center.)

Acceptor	Bond Length (pm)	Complex Ion	Bond Length (pm)
$CdCl_2$	223.5	$CdCl_6^{4-}$	253
GeF_4	167	GeF_6^{2-}	177
I_2	266	I_3^-	283
ICl	230	ICl_2^-	236
SO_2	143	SO_3^{2-}	150
$SbCl_5$	231	$SbCl_6^-$	247
SeO_2	161	SeO_3^{2-}	174
SiF_4	154	SiF_6^{2-}	171
SnI_4	264	SnI_6^{2-}	285
$ZrCl_4$	223	$ZrCl_6^{2-}$	245

GUYE'S RULE*

See Guldberg's rule.

HÄGG'S RULE

The basic rule concerning interstitial alloy formation is that of Hägg (1930):

> Below a critical radius ratio of 0.59, very simple metallic structures arise in which the non-metallic atoms can be considered to be inserted into the metal.

It is an essential feature of interstitial alloys that metallic properties, such as conductivity, are retained. As a result, elements which form ionic or homopolar bonding are excluded even though they obey Hägg's rule. In the case of Fe, borderline cases include Si, O, S, and P. These can form both compounds and interstitial alloys. See also the inverse Hägg's rule.

HAHN-FAJANS' RULE

> However greatly diluted, an element will be brought down in a crystalline precipitate if it can be built into the crystal lattice of the precipitate.

This is an improved version of Fajans' precipitation rule (qv). See also Paneth's adsorption rule.

HAHN-PANETH RULE

> An element is strongly adsorbed on a precipitate, even at very low concentrations, if the precipitate has a surface charge which is opposite to that

carried by the element and the adsorbed compound is only slightly soluble in the solvent.

This is an improved version of Paneth's adsorption rule (qv).

HAMMETT'S RULE

In the case of reactions of organic compounds which do not attack the nucleus directly, a rule has been established (Hammett, 1937) which relates the velocity constants and equilibrium constants. Thus, it links the values of the constants for a substituted compound to the values of the constants for an unsubstituted compound via a general logarithmic relationship which is such that the logarithm of the quotient can be expressed as the product of two factors. That is:

$$\log [k_s/k] = AB$$

where k_s is the constant for a substituted compound and k is the constant for an unsubstituted compound.

One of the factors, A, is a function only of the substituent and its position in the molecule, while B is a function only of the type of reaction. This means that the substituents are always arranged in the same series, regardless of the reaction involved, with regard to their effect upon reaction velocity or the position of equilibrium.

HAMMICK-ILLINGWORTH RULES

1. If, in a benzene derivative, C_6H_5XY, Y is in a higher group in the periodic table, then the second group to be added enters at the meta position.
2. If Y is in the same group of the periodic table but has a lower atomic weight than X, then again the second group to be added enters at the meta position.
3. In other cases, or when XY is a single group, the group is para-directing. When the groups joining X and Y are the same (e.g. -C = C-) the XY group is ortho-para orienting.

The rule can be strictly applied in the case of mixed groupings. For example, when adding $CHCl_2$, CH will be ortho-para orienting and CCl will be meta-directing.

HARDY-SCHULZE RULE

See the Schulze-Hardy rule.

HARKIN'S RULE

Atoms of even atomic number are more abundant in the universe than are atoms of odd atomic number.

HAYNES' RULE

Exciton binding to neutral donors and acceptors has been observed in many systems and is described by a relationship of the form:

$$E_b = A + BE_D$$

where E_b is the exciton binding energy, E_D is the donor binding energy, and A and B are constants.

The rule is remarkable in that it is often applicable far beyond the conditions for which it was originally derived.

HELICITY RULE

This rule (Charney, 1965) states that:

Skewed dienes exhibit an optical activity whose sign depends mainly upon the handedness of the helix which is formed by the 4 carbon atoms. A positive sign corresponds to a right-handed helix.

Its use usually gives good results for conjugated dienes which have nonpolar substituents in the immediate vicinity of the double bonds.

HERMANN'S RULE

This rule (Hermann, 1931) states that:

The energy which is absorbed in heating one gramme of an element from absolute zero to the melting point (including any allotropic changes), and then melting it, is equal to the melting point in degrees Kelvin, multiplied by a constant.

HERZ RULES

Various relationships between the properties of liquids and solids were explored (Herz, 1920, 1929a, 1929b), leading to the rules that:

$$a_{20} = 1/(2T_c - 293)$$

$$SL^{1/3} = \text{constant}$$

$$(L/c)^{1/3} = \text{constant}$$

where a_{20} is the coefficient of thermal expansion of a liquid at 20°, T_c is the critical temperature, S is the entropy of a solid at 25°, L is the latent heat of fusion, and c is the specific heat. The constants have a different value for different groups of elements.

HILDEBRAND'S RULE

When the density of the vapor phase above the liquid is 0.05mol/l, the change in entropy upon evaporating a mole of liquid is about 130J/mol.

The rule is not applicable to quantum liquids or to those which exhibit molecular association in the liquid phase.

HOFMANN RULE

If a quaternary ammonium compound contains an ethyl group, ethylene is formed in preference to any other olefin upon thermal decomposition.

If the compound contains other alkyl groups, including methyl, the latter remains attached to the N.

HOLLERAN-JESPERSEN RULES*

These authors suggested a simpler approach (Holleran & Jespersen, 1980) to the calculation of oxidation numbers. This was in the form of a number of hierarchical rules:

1. The sum of the oxidation numbers of all of the atoms in a molecule or ion must equal the total charge. This is equal to zero for neutral molecules.

2. The metallic atoms of group 1A have an oxidation number of + 1, and those of group 2A have an oxidation number of + 2.
3. Hydrogen is assigned an oxidation number of + 1 and F is assigned an oxidation number of − 1.
4. Oxygen is assigned an oxidation number of − 2.
5. Atoms of group 7A are assigned an oxidation number of − 1.
6. In binary compounds, atoms from group VIA are assigned an oxidation number of − 2 and atoms from group VA are assigned an oxidation number of − 3.

If two rules conflict, the rule which is earlier in the list takes precedence. If a conflict occurs within the same group, the lighter element follows the rule. If a molecule or ion contains one element which is not mentioned in these rules, its oxidation number is found by difference.

HONDROS' RULE*

When carrying out certain metallurgical or chemical processes, it is often useful to have some guide as to which of the constituents in solution will be enriched or depleted at the free surface. Hondros (1980) made a theoretical study of the popular assumptions, that the element with the lowest surface tension or melting point will enrich the surface, and deduced the rule that:

> In any binary liquid system, the component which enriches the surface is that which has the lower surface tension. The greater the difference in surface tension between the two pure components, the greater will be the surface enrichment of the component having the lower surface tension.

He suggests that the first (qualitative) part of the rule is always true, while the second (quantitative) part is only generally true.

HOPPER'S RULE*

> Electricity travels one foot in a nanosecond.

One of a number of rules (Bentley, 1985) compiled for the convenience of computer programmers. Not a rule to interest the average computer user, but very useful if one is rewiring a Cray-3.

HOVLAND'S RULES*

The writing of the ground state electron configurations of the elements, according to the aufbau principle, is a standard exercise. The configurations are also the starting point for some of the other rules in this compilation. Various methods have been proposed for remembering these configurations. Hovland (1986) suggested a rule which had the advantage of being based upon the familiar form of a chessboard (Figure 34). The rules for filling the black squares of the board are:

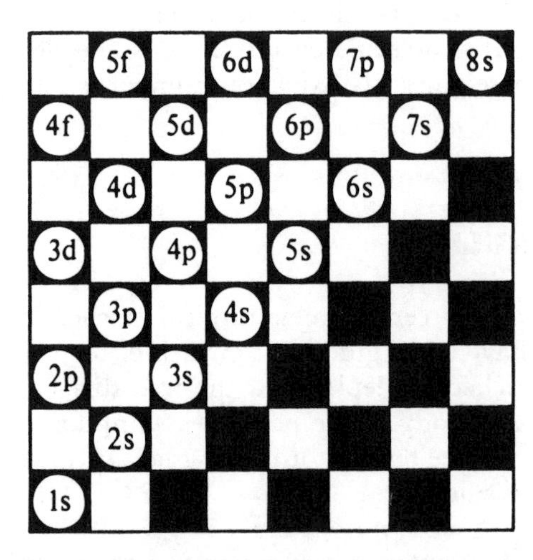

Figure 34. The Hovland rule for writing the ground state electron configurations of the elements. The configurations are read from left to right, moving up the board.

1. The squares on the diagonal running from lower left to upper right are marked with an "s" and are numbered from 1 to 8; starting from the lower left.
2. Moving up to the next diagonal of the same color, the squares are marked "p" and are numbered from 2 to 7; starting from the lower left.
3. The third and fourth diagonals are successively labelled "d" and "f" and are numbered from 3 to 6, and from 4 to 5, respectively; starting from the lower left.

The electron configuration for a given element is then obtained by starting at the lower left and reading from left to right on successive rows. The first number of each diagonal corresponds to the first principal quantum level which contains the associated sub-shells. That is, the p sub-shells for which n must be equal to 2 or more begin on the second diagonal; and so on. Among the advantages suggested for this method are that one reads naturally from left to right, and that the chessboard is familiar. Unfortunately, the latter fact can cause confusion, and Hovland himself used an incorrectly oriented chessboard. See also the Yi rule and the Carpenter rule.

HSU'S RULE*

Having estimated the critical driving force associated with martensitic transformations, Hsu (1985) went on to deduce that the martensite start temperature, M_s of an Fe-C alloy (plain carbon steel) is related to the composition by:

$$M_s = 520 - 320C$$

where M_s is expressed in degrees Celsius and C is the carbon content (%).

HÜCKEL'S RULE

The original form of this rule states that:

> Planar organic ring systems involving equal bond lengths are especially stable when the number of pi-electrons is of the form, $4n + 2$.

It has been shown subsequently that the conditions of planarity and equality of bond length can be relaxed in some cases.

HUDSON'S RULES

On the basis of data concerning the optical rotatory properties of a wide range of substances, but especially pyranose sugars and lactones, Hudson (1909, 1910, 1939) and others (Levene, 1915) formulated a number of rules:

1. The rotatory contribution of the carbon-1 (glucoside) atom is affected only slightly by changes in the structure of the remainder of the molecule and, in the D series, the more dextrorotatory anomer is always the alpha-form.

2. A gamma-lactone in which the oxide ring lies to the right of the conventional projection formula is more dextrorotatory than the parent acid. If the ring lies to the left, the lactone is more laevorotatory.[1]
3. If the rotation which is contributed by carbon-1 is called A, and that contributed by the remaining asymmetric centers is called B, then the molecular rotations will be A + B for the alpha-form of the d-series and B − A for the beta-form of the d-series.[2]
4. Changes at the carbon-1 atom affect only slightly the rotation of the remainder of the molecule.
5. The sign of the rotation of the phenylhydrazides of the sugar acids is determined by the sign of carbon-2. When this is (+), the hydrazide is dextrorotatory. When it is (−), the hydrazide is laevorotatory.[3]
6. All alpha-hydroxy acids whose rotation is shifted to the right by their transformation into amides have the configuration[4]:

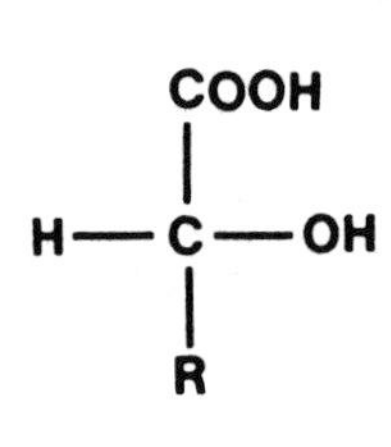

HUME-ROTHERY RULES

There are a number of rules which are habitually included under this heading. For instance the B sub-group elements, in which the outer shells are partially filled, are usually less closely packed and the bind-

---

[1] *This is also known as the lactone rule. The significance of the position of the oxide (lactone) ring is that it reveals the former position of the -OH group on the gamma-C atom. The rule cannot be directly applied to a ketose, rather than an aldose, because the former do not give acids and lactones without breaking up.*

[2] *These are also known as the rules of isorotation. It follows that, in an alpha/beta- pair of isomers, the sum of their molecular rotations will be a constant which is characteristic of the particular sugar and the difference will be a constant which is characteristic of the nature of the hydroxyl group or substituted hydroxyl on carbon-1.*

[3] *This is also known as the hydrazide rule. It also applies to sugar amides and to the acetylated nitriles of sugar acids.*

[4] *This is also known as the amide rule.*

ing is more covalent. Most of the structures of elements in the B subgroups obey the 8-N rule which states that:

> Each atom has 8-N close neighbors, where N is the number of the group to which the element belongs.

The rule does not apply to group IIIB elements, among others. Thus, Ga does not have the expected coordination number of 5. It is natural that extensive solid solutions should not be formed when the sizes of the solute and solvent atoms are very different. This was quantitively expressed by the rule that:

> If the atomic diameters of two metals differ by more than about 14%, the primary solid solutions are usually restricted to a few atomic percent.

In the opposite case, extensive solid solubility cannot necessarily be expected. The percentage was later increased to 15% and this is therefore sometimes called the 15% rule. It was also pointed out that some fundamental similarities become apparent between alloy systems if the number of valence electrons per atom is computed for each alloy. The rule can be stated as:

> In alloys of group IB elements with IIB elements, analogous beta-, gamma-, and epsilon-phases are found at electron/atom ratios of 3:2, 21:13, and 7:4, respectively.

A final rule which belongs in this group is the one which states that:

> Factors which cause a marked decrease in the liquidus slope have a greater relative effect upon the solidus slope.

HUND'S RULES

The general problem of assessing the relative energies of all of the possible states of the electron configuration of an atom is very difficult. However, these empirical rules indicate which will be the most stable state (the ground state). The rules (Hund, 1925) suggest that:

1. The ground state will always have the highest value of spin multiplicity.
2. If several states have the highest spin multiplicity, the most stable one will be that which has the highest value of the quantum number characterising the resultant orbital momentum.

3. The energies of the sub-states will increase, as the total angular momentum increases, for a state which is derived from a configuration with a shell which is less than half full. The order of the sub-states is reversed when the shell is more than half full.

It must be emphasized that these rules are generally applicable only to the ground state. The first rule is sometimes facetiously known as the "public transport" principle. This is because psychologists have noted that strangers who are boarding public transport will not sit together until all of the double seats are each occupied by a lone person. In the same way, the first rule implies that two electrons (of opposite spin) will not occupy the same energy level until all of the available levels each have one electron.

HUTCHINGS' RULE*

It has been found (Hutchings, 1975) that, for pure metals:

> The erosion resistance of a metal is proportional to the product of the specific heat, the density, and the difference between the melting point and the test temperature.

See also the rules of Ascarelli, Brauer & Kriegel, Finnie-Wolak-Kabil, Khruschov, Smeltzer, and Vijh.

HYDRAZIDE RULE

See Hudson's rules.

ICE RULES

See the Bernal-Fowler rules.

INTENSITY SUM RULES

1. The sum of the intensities of all of the lines of a multiplet which start from a common initial level is proportional to the quantum weight, $2J+1$, of the initial level.
2. The sum of the intensities of all of the lines of a multiplet which end on a common final level is proportional to the quantum weight, $2J+1$, of the final level.

INVERSE HÄGG RULE

The Hägg rule (qv) for interstitial alloy formation defines the critical radius ratio below which the smaller element can form an interstitial alloy with a metal. The metal in question is usually Fe. In certain compound-like alloys, such as FeSb, the other element is so large that Fe plays the interstitial role. That is, the same radius ratio applies but the positions of the two components are inverted.

ISOPRENE RULE

Natural substances in which the number of C atoms is an even multiple of 5 can be considered to be made up of isoprene structures.

Terpenes (C_{10}), sesquiterpenes (C_{15}), and diterpenes (C_{20}) are common constituents of volatile oils, and triterpenes (C_{30}) and tetraterpenes (C_{40}) are known. However, isoprene itself cannot be considered to be a natural product.

ISO-ROTATION RULES

See Hudson's rules.

KAPELLOS-MAVRIDES RULES*

The concept of electronegativity is useful in every branch of chemistry, and is also a relatively fundamental quantity to which many other properties can be related using simple correlations. It was pointed out only recently (Kapellos & Mavrides, 1987) that:

1. The electronegativities (on Pauling's scale) of the second-row elements of the periodic table are equal to:

   $(Z - 1)/2$

2. The electronegativities (on Pauling's scale) of the third-row elements of the periodic table are equal to:
   $(Z^* - 1)/3$

where Z is the atomic number and $Z^* = Z - 7$.

In the first case, the values are correct within experimental error. In the second case, the value is exact for Cl but is 0.1 or 0.2 units too high for all of the other elements in the row (appendix 3).

KATZ RULES*

In designing with ceramics for high-performance structural applications, codes of practice and experience are often lacking. As such design is more of an art than a science, Katz (1985) suggested some tentative rules:

1. Avoid point loads. Areal loading is best, followed by line loading.
2. Maintain structural compliance. Use compliant layers, and parts with mating radii.
3. Avoid stress concentrators. Avoid or minimize sharp corners, rapid changes in section size, undercuts, and holes.
4. Limit the effect of thermal stresses: minimize the section size and maximize the degree of symmetry. The greater the symmetry the smaller is the influence of thermal stresses. Replace complex components by simpler parts having higher symmetry.
5. Keep components as small as possible. The flaw distribution of ceramics causes the strength to be size-dependent. Minimizing the component size increases the reliability.
6. Minimize the severity of impact. Where impact, including particulate erosion, cannot be avoided, design so as to obtain low-angle (20 to 30°) impacts.
7. Machine the components very carefully. The most critical strength-reducing flaws are often surface or near-surface cracks which were introduced during grinding or other machining processes.

KAUFFMAN'S RULES*

It was pointed out (Kauffman, 1986) that the accepted rules for the determination of oxidation states are difficult to apply to atoms which are covalently bound in compounds. It was instead suggested that one should use a table of electronegativity values (appendix 3), together with the following rules:

1. Draw a Lewis structure for the compound or ion, showing all unshared pairs of electrons (where equivalent resonance structures exist, only one need be drawn).
2. Re-draw the structure in "exploded" form, so that the more electronegative atom at the end of each bond retains all of the electrons which

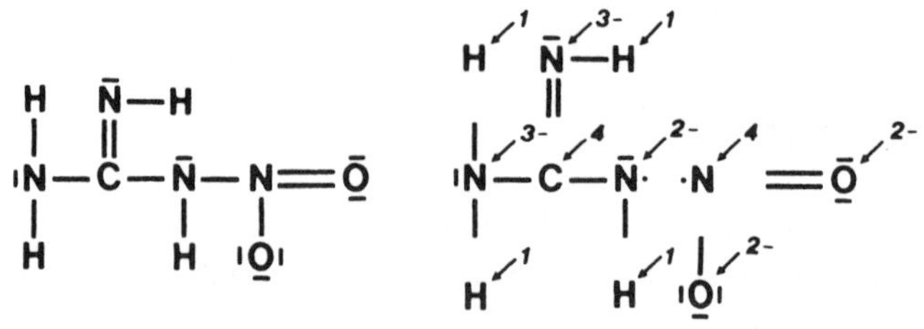

Figure 35. The use of Kauffman's rule to determine the oxidation numbers of each atom in the covalent compound, nitroguanidine. First draw the Lewis structure (a), then "explode" it (b) and apply rule 3. The arrows indicate the oxidation numbers deduced for each atom.

had been bonding electrons. If a bond is between atoms of the same element, give half of the electrons to each atom. Each atom retains its own non-bonding electrons.

3. Subtract the number of electrons around each separate atom from the number of valence electrons which the free atom would have. This difference is the oxidation number.

In Figure 35, this method is applied to nitroguanidine.

KEMPSTER-LIPSON RULE*

The number of C, N, and O atoms in the unit cell of an organic crystal is approximately equal to the volume of the cell (in cubic Angstrom units), divided by 18.

Assuming that the size of the unit cell has been deduced, by using X-ray diffraction for example, this rule (Kempster & Lipson, 1972) is usefully applied (instead of measuring the density) when deciding whether it is feasible to determine the structure.

KERESELIDZE-KIKIANI RULES*

These are among a number of simple correlation rules which relate the energy levels in separate atoms to those in a molecule. The present authors proved three such rules (Kereselidze & Kikiani, 1984), but the first of these was the same as that proposed by Barat and Lichten (qv). The other rules state that:

$n_{u,a} = 2n_{i,a}$, for symmetrical orbitals

$n_{u,a} = 2n_{i,a} + 1$, for antisymmetrical orbitals

where $n_{u,a}$ is the number of zeroes of the angular wave function for the united atom and $n_{i,a}$ is the number of zeroes of the angular wave function for the isolated atom.

See also the Eichler-Wille rule.

KERN'S RULES

These rules were suggested (Kern, 1950) for the estimation of the thermal conductivities of liquid mixtures, solutions, and dispersions. They state that:

1. For mixtures of organic liquids, use an average conductivity, weighted by the mass fractions.
2. For mixtures of organic liquids with water, take 90% of the mass-weighted conductivity.
3. For solutions of salts in water, take 85% of the conductivity of water; up to a concentration of 30% by weight.
4. For colloidal dispersions, take 90% of the conductivity of the dispersion liquid.
5. For emulsions, take 90% of the conductivity of the continuous liquid phase surrounding the droplets.

KEYES' RULES*

The best known rule to be given this name (Keyes, 1962) states that, for a series of homologous tetrahedrally bonded semiconductors,

$$C_{ij}d^4 = \text{constant}$$

where C_{ij} is an elastic constant, and d is the nearest-neighbor distance. Another rule relates the activation volume for diffusion, V, to the activation energy for diffusion, E, and the compressibility, K:

$$V = 4KE$$

Because it does not permit negative values of the activation energy, this rule is somewhat dubious.

KHRUSCHOV'S RULES*

It has been found (Khruschov, 1974) that, for pure metals, the abrasive wear resistance can be estimated from the rules:

$$e/H = \text{constant}$$

$$e/E^{4/3} = \text{constant}$$

$$e/W^2 = \text{constant}$$

where e is the wear resistance, H is the hardness, E is the elastic modulus, and W is the heat content.

The latter quantity is found by adding the heat of fusion to the product of the density, the specific heat, and the temperature rise required to cause melting (see Hutchings' rule). See also the rules of Ascarelli, Brauer & Kriegel, Finnie-Wolak-Kabil, Smeltzer, and Vijh.

KIER'S RULE*

Sweetness is a subjective property which is detected only by receptors on the tongue. That is, no instrument yet devised can measure this commercially important feature. However, it has been realized that many of the chemicals which are already known for their sweetness have a common characteristic. This is known as the triangle of sweetness, and one can summarize the concept by the rule that:

> **Sweetness is associated with the presence, somewhere in a molecule, of a triangle made up of two H-bonding groups and one H-repelling group.**

This statement is insufficient without an exact specification of the size of the triangle and the nature of the various groups. The basic triangle is shown schematically in Figure 36. Here, AH is an atom of O or N which carries an H atom, B is an atom capable of forming a hydrogen bond, and X is a group which repels H. The AH and B corners are often amide and carbonyl groups. The two dimensions indicated are critical and are believed to be related to the size of the receptors on the tongue. It should be clearly understood that the triangle is an imaginary one and that the sides of the triangle are not chemical bonds (Figure 37). Moreover, there are sweet substances to which this triangle rule seems completely inapplicable. For instance, the metal now known as beryllium

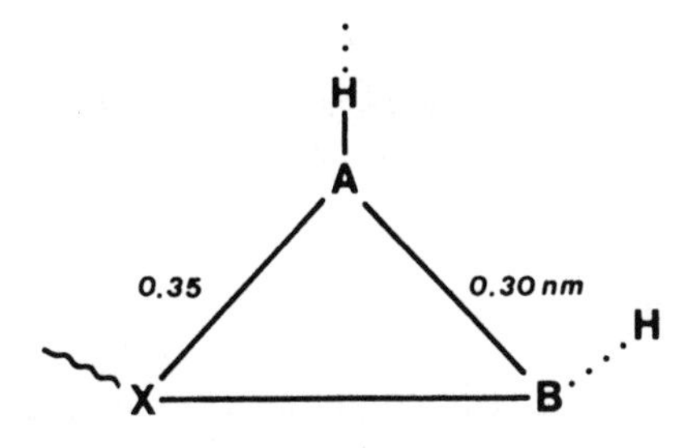

Figure 36. The triangle or Kier rule for sweetness. Here, AH is a group carrying an H atom which can form H-bonds, B is also a group which can form H-bonds, and X is a H-repelling group. The distances of 0.30 and 0.35nm are critically important but the sides of the triangle do not represent chemical bonds.

was originally called glucinium (see older textbooks) because many of its salts are sweet. Yet another possible reason for the fall of the Roman Empire may be that they added lead to wine and grape juice. The lead salts thus formed were sweeteners.

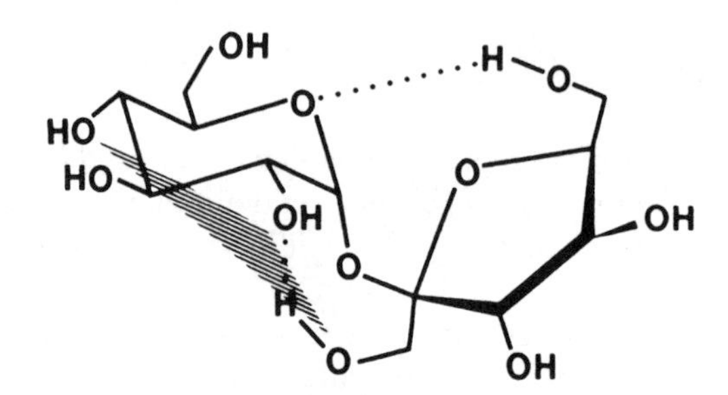

Figure 37. The triangle or Kier rule for sweetness, as applied to sucrose. Note that the molecule consists of a glucose ring and a fructose ring and that the constituent groups of the triangle are from different rings. That is, the sides of the triangle do not represent chemical bonds.

KIREEV'S RULE

This rule (Kireev, 1929) states that:

Trouton's rule (qv) applies to each component of mixtures of normal liquids if the ratio of the partial pressure to the fractional amount is a constant.

KOHLER'S RULE

In the study of magnetoresistance and related phenomena, an important rule is that:

The increase in resistivity due to a magnetic field, divided by the zero-field resistivity, is a universal function of the magnetic field strength, divided by the zero-field resistivity.

KOLB'S RULES*

These were proposed (Kolb, 1978) for the calculation of the oxidation number of a substance:

1. For any free (uncombined) element, the oxidation number is zero.
2. For any neutral compound, the sum of the oxidation numbers is zero.

3. For any monatomic ion, the oxidation number is equal to the ionic charge.
4. For any polyatomic ion, the sum of the oxidation numbers is equal to the ionic charge.
5. In the case of H, the oxidation number in compounds is + 1; except for metal hydrides, where it is − 1.
6. In the case of O, the oxidation number in compounds is − 2; except for peroxides, where it is − 1. In F_2O, oxygen has an oxidation number of + 2 since F is the only element more electronegative than O.
7. In compounds, group IA elements are always + 1, group IIA elements are always + 2, and group IIIA elements are almost always + 3.

KONOWALOFF'S RULE

The vapor of a mixture of two liquids is richer in the component whose addition to the mixture leads to an increase in the total pressure.

KOO-COHEN RULES*

An empirical rule has been suggested (Koo & Cohen, 1987) for the calculation of the single crystal elastic constants of a binary alloy between a cubic solvent and an hexagonal solute:

The elastic constants of a single crystal of an alloy between a cubic solvent and a hexagonal solute are given by:

$$C_{11} = xC_{11}' + (1 - x)C_{11}''$$

$$C_{12} = xC_{12}' + (1 - x)C_{12}''$$

$$C_{44} = xC_{44}' + (1 - x)C_{44}''$$

where x is the atomic fraction of the hexagonal solute, the C_{ij} are the elastic constants of the alloy, the C_{ij}'' are the elastic constants of the cubic solvent, and the C_{ij}' are the average elastic constants of the hexagonal solute:

$$C_{11}' = (2C_{11} + C_{33})/3$$

$$C_{12}' = (C_{12} + 2C_{13})/3$$

$$C_{44}' = (2C_{44} + C_{66})/3$$

This method seems very complicated at first sight, but the basic relationships can be regarded as a simple rule of mixtures and the apparently illogical expressions used in the averaging of the hexagonal constants can be appreciated immediately upon superposing the symmetrical matrices of the elastic constants for the cubic and hexagonal crystal structures (Figure 38). The rule gives good results for so-called normal alloys but does not apply to alloys which have anomalous phonon properties or exhibit an anomalous behavior in their elastic constants.

$$\begin{matrix} C_{11} & C_{12} & C_{12} & 0 & 0 & 0 \\ & C_{11} & C_{12} & 0 & 0 & 0 \\ & & C_{11} & 0 & 0 & 0 \\ & & & C_{44} & 0 & 0 \\ & & & & C_{44} & 0 \\ & & & & & C_{44} \end{matrix}$$

$$\begin{matrix} C_{11} & C_{12} & C_{13} & 0 & 0 & 0 \\ & C_{11} & C_{13} & 0 & 0 & 0 \\ & & C_{33} & 0 & 0 & 0 \\ & & & C_{44} & 0 & 0 \\ & & & & C_{44} & 0 \\ & & & & & C_{66} \end{matrix}$$

Figure 38. Matrix of Elastic Constants for Cubic and Hexagonal Systems (Upper matrix: cubic system, lower matrix: hexagonal system)

KOPP'S RULE

This rule (Kopp, 1842) suggests that:

Equal differences in the chemical composition of organic compounds give rise to equal changes in boiling point.

Thus, the ethyl ester of an acid boils at a temperature which is an average of 20° higher than that for the methyl ester.

KOPP-NEUMANN RULE

See Neumann-Kopp rule.

KOSO RULE

The high-temperature hardness of a material is an important parameter in itself, and can also be a useful guide to the magnitudes of other properties. Unfortunately, the hot hardness can be difficult to measure. A notable feature of a hardness versus temperature graph (on log-log coordinates) is an inflection point which occurs at a certain temperature, T_t. It is found (Kagawa et al., 1984) that, for a wide range of materials, one can use the rule:

$$H_t = 0.7H_r$$

where H_t is the hardness at the inflection temperature and H_r is the room temperature hardness.

KUBASCHEWSKI'S RULES*

In order to calculate thermochemical equilibria, a knowledge of the standard entropies and of the heat capacities of double molecules in the gas phase is sometimes required. However, only a few of these data have been measured experimentally. Recognizing this, Kubaschewski (1984) presented empirical rules which permit a reasonably accurate estimation of the entropies and atomic heats of larger molecules. These were based upon available data (Tables 12 and 13) and state that:

Table 12
Standard Entropies of Gaseous Molecules
(see Kubaschewski's rules)

Molecule	Entropy (cal/molK) Single	Double	Ratio
Al_2Br_6	83.43	130.8	1.57
Al_2Cl_6	75.12	113.65	1.51
Al_2F_6	66.4	92.5	1.39
Al_2I_6	86.78	139.6	1.60
Be_2Cl_4	60.0	91.15	1.52
Co_2Cl_4	71.3	107.6	1.51
Fe_2Br_4	80.6	123.3	1.53
Fe_2Cl_6	82.24	128.35	1.56
Fe_2I_4	83.5	129.9	1.56
K_2Br_2	59.85	89.9	1.50
K_2I_2	61.7	94.6	1.55
Li_2Br_2	53.6	75.15	1.40
Li_2I_2	55.48	79.0	1.43
Na_2Br_2	57.63	83.38	1.44
Ti_2Cl_6	75.7	115.0	1.52
W_2Cl_{10}	96.9	170.0	1.75

1. The ratio of the entropy of a double gaseous molecule at 298K, to that of a single gaseous molecule at 298K, is equal to 1.52.
2. The ratio of the heat capacity of a double gaseous molecule to that of a single gaseous molecule is equal to 2.16.

He points out that there is an astonishingly small scatter of the data around these ratios, and that the rules can be particularly recommended.

Table 13
Atomic Heats of Gaseous Molecules
(see Kubaschewski's rules)

		Atomic Heat (cal/molK)		
Molecule	**Temperature (K)**	**Single**	**Double**	**Ratio**
Al_2Br_6	300	18.04	39.9	2.22
Al_2Br_6	500	19.1	42.24	2.21
Al_2Cl_6	300	17.0	37.9	2.23
Al_2Cl_6	500	18.7	41.4	2.22
Al_2F_6	300	14.98	32.05	2.14
Al_2F_6	500	17.23	38.0	2.20
Al_2I_6	300	18.28	40.9	2.52
Al_2I_6	500	19.26	42.65	2.39
Be_2Cl_4	300	12.27	27.53	2.24
Be_2Cl_4	500	13.64	30.0	2.20
Co_2Cl_4	300	14.26	30.42	2.14
Co_2Cl_4	500	14.66	31.42	2.15
Fe_2Br_4	300	14.53	31.3	2.13
Fe_2Br_4	500	14.68	31.5	2.15
Fe_2Cl_6	300	18.6	41.53	2.23
Fe_2Cl_6	500	19.4	42.9	2.21
Fe_2I_4	300	14.58	31.42	2.15
Fe_2I_4	500	14.71	31.56	2.14
K_2Br_2	300	8.84	19.7	2.23
K_2Br_2	500	8.98	19.72	2.19
K_2I_2	300	8.9	19.8	2.21
K_2I_2	500	9.0	19.8	2.21
Li_2Br_2	300	8.1	17.94	2.21
Li_2Br_2	500	8.66	19.15	2.21
Li_2I_2	300	8.26	18.18	2.20
Li_2I_2	500	8.74	19.24	2.20
Na_2Br_2	300	8.84	19.47	2.20

Table 13 (continued)
Atomic Heats of Gaseous Molecules

Molecule	Temperature (K)	Atomic Heat (cal/molK) Single	Double	Ratio
Na_2Br_2	500	8.9	19.53	2.19
Ti_2Cl_6	300	17.37	33.96	2.23
W_2Cl_{10}	300	28.77	62.98	2.19
W_2Cl_{10}	500	30.58	65.8	2.15

KUCZYNSKI'S RULE *

Using simple arguments, theoretical relationships can be obtained between the elastic constants and various other physical properties and thermodynamic data. In the case of metals:

$$Gd/E = \text{constant}$$

$$G = L/V$$

where G is the shear modulus, d is the interatomic spacing, E is the surface energy, L is the sublimation energy, and V is the atomic volume.

The value of the constant is theoretically equal to 8. The value found on the basis of experimental data for common metals is about 9. See also the Rice-Thomson rule.

KUHN-THOMAS-REICHE RULE

For an atom whose electrons are undergoing all of the possible transitions to or from a level, L_2, this rule states that:

$$\Sigma f_{12} + \Sigma f_{23} = Z$$

where L_1 and L_3 are all of the levels below and above L_2, respectively, the f_{ij} are the oscillator strengths for each transition, and Z is the number of optical electrons.

KUMLER'S RULE

It was found (Kumler, 1935) that:

$$(e - n^2)^{1/2}/m = \text{constant}$$

where e is the dielectric constant, n is the refractive index, and m is the dipole moment. This applies to a wide range of liquids.

KUNDT'S RULE

This states that:

The refractive index of a medium on the shorter wavelength side of an absorption band is abnormally low, and that on the longer wavelength side is abnormally high.

KÜSTER'S RULE

It was suggested (Küster, 1890), on the basis of studies of organic binary phase diagrams, that:

The solidifying point of an isomorphous mixture lies on a straight line joining the melting points of the two components.

This rule obviously has many exceptions.

LACTONE RULE

See Hudson's rules.

LAPORTE'S RULE

In dipole radiation, only transitions between terms of opposite parity are allowed.

That is, transitions between terms having either even S or odd S are allowed, where $S = \Sigma l_n$ is the sum of the absolute values of the quantum number, l, for all of the electrons.

LAPWORTH'S RULE

This is an extension (Lapworth, 1898), of the concept of the alternation of affinity, which proposes that alternate atoms in an organic compound acquire similar polarities and similar reactivities. It states that:

In intramolecular changes of an organic compound, the labile group moves from an alpha-atom and attaches itself to a gamma-atom.

An alternative and more general statement is that:

A labile group moves along a chain of alternately singly and doubly-bound atoms with the ethylenic and single linkings changing pairs in the path of the labile group.

LAUGIER'S RULE *

Empirical relationships between the hardness and other properties of the cemented carbide, WC-Co, are important in quality control. One of the simplest and most useful rules (Laugier, 1985) is:

$$H/H_c^{1/3} = \text{constant}$$

where H is the hardness and H_c is the coercivity. When H is expressed in kg/mm^2 and H_c is expressed in kA/m, the numerical value of the constant is 600.

LE CLAIRE'S RULES

It has been proposed (Le Claire, 1976) that normal self-diffusion in metals obeys the rules that:

1. The diffusion coefficient obeys the Arrhenius law.
2. The pre-exponential values range from 5×10^{-6} to $5 \times 10^{-4} m^2/s$.
3. $E/T_m = 34$

In the last rule, E is the activation energy (cal/mol) and T_m is the melting point (K) of the metal. This is just a quantitative statement of the Van Liempt rule (qv). In practice, the operation of mechanisms other than the monovacancy one may cause deviations from the rules, especially the first one. See Shewmon's rules and Tiwari's rule.

LEE-KIM RULE *

Combining rules are often used (Stwalley, 1971) to estimate the interaction potential, P_{AB}, of asymmetric pairs of atoms or molecules when those for the symmetric pairs (P_{AA}, P_{BB}) are known. The most commonly used rules are the arithmetic mean combining rule (qv) and the geometric mean combining rule (qv). However, these rules have disadvantages when the difference in the sizes of A and B is large. This is because the interactions in all three systems are assumed to occur at equal values of separation, R. The Smith rule (qv) for the repulsive interaction incorporates into the arithmetic mean the fact that the two atoms or ions in the AB pair are different. The method involves a calculation of the distance between nucleus A and an imaginary boundary, between the two species, which is located at r_A. Lee and Kim (1981) modified the Smith rule by assuming that r_A/R is equal to $p_A/(p_A + p_B)$; where the p are the Pauling ionic radii (appendix 3) of atoms A and B. Thus:

The asymmetric pair interaction potential between two atoms or ions is given by:

$$P_{AB}(R) = [P_{AA}(2z_AR) + P_{BB}(2z_BR)]/2$$

where $z_A = p_A/(p_A + p_B)$, $z_B = p_B/(p_A + p_B)$, the p values are Pauling ionic radii, and $P_{AA}(2z_AR)$ is the A-A interaction potential at separation, $2z_AR$, etc.

LEE-ROBERTSON-BIRNBAUM RULES*

In the field of mechanical properties, an important problem is to predict which slip system will be activated in a second grain by a dislocation pile-up in a first grain. The present rules (Lee et al, 1989) make prediction easier, and suggest that:

1. The angle between the traces of the slip planes on the grain boundary plane should be a minimum.
2. The magnitude of the Burgers vector of the residual dislocation left in the grain boundary should be a minimum.
3. The resolved shear stress acting on the slip system should be a maximum.

In order to predict the slip system, rules 2 and 3 have to be considered together. The dominant system should leave a dislocation with a small residual Burgers vector in the boundary, and have a reasonable resolved shear stress acting upon it.

LEWIS STRUCTURE RULES

These are rules which suggest the steps to be followed in the writing of Lewis formulae:

1. Decide whether the bonds are ionic, covalent, or both.
2. Write down the skeleton of the compound.
3. Calculate the total number of valence electrons and put them into the skeleton.

This system is surprisingly reliable, although using the rules without taking account of other information can lead to incorrect conclusions;

even in the case of compounds as simple as the oxygen molecule. There are also a large number of cases (including CO_2) where more than one permissible Lewis structure can be drawn but where only one compound exists.

LEWIS-RANDALL RULE

The fugacity of a constituent in a mixture of gases at a given temperature is proportional to its mole fraction.

This rule is regarded as being an improvement on Dalton's law of partial pressures.

LINDE'S RULE

This states that the general expression for the increase in resistivity per atomic percent of substitutional impurity added to a monovalent metal is:

$$dR = a + bZ^2$$

where a and b are constant for a given row of the periodic table and for a given solvent metal, and Z is the excess valency.

LITTLE'S RULE*

The average number of objects in a queue is equal to the product of the entry rate and the average holding time.

One of a number of rules (Bentley, 1985) compiled for the convenience of computer programmers. Impress your fellow customers while awaiting service in a single-queue bank.

LIVINGSTONE'S RULE

According to this rule (Livingstone, 1936), the change in boiling point due to a change in pressure is given by:

$$dT = T_b dP/10$$

where dT is the change in boiling point, T_b is the normal boiling point, and dP is the change in pressure in mm. This rule over-corrects the boiling points of associated liquids.

LOBO-MILLS RULE*

On the basis of tracer diffusion results for various solutes in Hg (Lobo & Mills, 1982), it was proposed that:

In mercury amalgams at room temperature, the diffusivity of a solute atom is inversely proportional to its Goldschmidt atomic radius.

Of course, such correlations between diffusivity and atomic size are often reported. For instance, an inverse relationship between solid state diffusivity and Pauling atomic radius has been found for stainless steel (Patil et al, 1980). This particular rule for Hg is given only because liquid Hg tends to play an important role in many different scientific disciplines.

LONGCHAMBON RULE

It was proposed (Longchambon, 1921) that:

The optical rotatory power can be greater or less in the crystalline state than it is in the liquid or fused state. However, it is usually greater.

LONGINESCU'S RULE

It was proposed (Longinescu, 1903) that, for organic compounds:

$$T_m/dn^{1/2} = \text{constant}$$

where T_m is the melting point, d is the density at 0°C, and n is the number of atoms in the molecule. It was also suggested that:

$$T_b M N n^{1/2} = \text{constant}$$

where T_b is the boiling point, M is the molecular weight, and N is the number of molecules per unit volume.

LORENTZ RULES

See the arithmetic combining rule.

LORENZ RULE*

$$T_c = 0.9(T_m + T_b)$$

where T_c is the critical temperature, T_m is the melting point, and T_b is the boiling point.

The author (Lorenz, 1916) derived this expression from Guldberg's rule (qv). See also Prudhomme's rule, Porlezza's rule, and the Do-Yen-Chen rule.

LUZZI-MESHII RULES*

The transformation of some intermetallic compounds from the crystalline to the amorphous state can be achieved by means of electron irradiation. In order to help to distinguish between those substances which can be so transformed and those which cannot, a number of criteria have been established (Luzzi & Meshii, 1986). The present rules state that an intermetallic compound will amorphize under electron irradiation if:

1. The critical temperature for chemical ordering in the absence of irradiation is higher than its melting point.
2. Both of its constituents are present in concentrations which are equal to, or greater than, 1:3.
3. The constituents are separated by more than two groups in the periodic table.
4. The material has a complex crystal structure in the irradiated state.

Note that these are not alternative possibilities. All of them must be obeyed simultaneously (Table 14). See also Ziemann's rule.

Table 14
Amorphization and Other Characteristics of Intermetallic Compounds
(Underlined data favor amorphization according to the Luzzi-Meshii rules)

Compound	Structure	C-A[1]	T[2]	f	dN
CoTi	B2	no	<u>no</u>	<u>0.50</u>	<u>5</u>
Co_2Ti	<u>C15</u>	yes	<u>no</u>	<u>0.33</u>	<u>5</u>
Cr_2Ti	<u>C15</u>	no	<u>no</u>	<u>0.33</u>	2
Cu_3Au	$L1_2$	no	yes	0.25	0
$CuTi_2$	<u>$C11_b$</u>	yes	<u>no</u>	<u>0.33</u>	<u>7</u>
CuTi	<u>B11</u>	yes	<u>no</u>	<u>0.50</u>	<u>7</u>
Cu_4Ti_3	<u>Cu_4Ti_3</u>	yes	<u>no</u>	<u>0.43</u>	<u>7</u>

Cu_3Ti_2	Al_3Os_2	yes	no	0.40	7
Cu_4Ti	Au_4Zr	no	no	0.20	7
CuZn	B2	no	yes	0.50	1
CuZr	CuZr	yes	no	0.50	7
$Cu_{10}Zr_7$	$Ni_{10}Zr_7$	yes	no	0.41	7
FeAl	B2	no	no	0.50	5
FeCo	B2	no	yes	0.50	1
$FeNi_3$	$L1_2$	no	no	0.25	2
FeNi	$L1_0$	no	yes	0.50	2
FeTi	B2	no	no	0.50	4
Fe_2Ti	C14	yes	no	0.33	4
Mn_2Ti	C14	yes	no	0.33	3
Nb_7Ni_6	$D8_5$	yes	no	0.46	5
NiAl	B2	no	yes	0.50	3
Ni_3Al	$L1_2$	no	no	0.25	3
Ni_3Mn	$L1_2$	no	no	0.25	3
NiMo	orthorh.	yes	no	0.50	4
Ni_3Mo	orthorh.	no	no	0.25	4
Ni_4Mo	$D1_a$	no	no	0.25	4
$NiTi_2$	$E9_3$	yes	no	0.33	6
NiTi	B2	yes	no	0.50	6
Ni_3Ti	$D0_{24}$	no	no	0.25	6
Zr_2Al	$B8_2$	yes	no	0.33	9
Zr_3Al	$L1_2$	no	no	0.25	9
Zr_2Ni	C16	yes	no	0.33	6

[1] *Crystalline to amorphous transition.*

[2] *Order/disorder transition at high temperatures in the absence of irradiation (rule 1). Here, f is the fractional composition of the minor constituent (rule 2) and dN is the number of groups between the two constituents (rule 3).*

MACKENZIE-LICHTENBERGER RULE*

A method which is often used to estimate the vacancy formation energy from positron annihilation data is to use the rule (MacKenzie & Lichtenberger, 1976) that:

$E_f/T_t = 15k$
where E_f is the vacancy formation energy, T_t is the threshold temperature for positron trapping, and k is Boltzmann's constant.

MARCH-RICHARDSON-TOSI RULE*

These authors (March et al, 1980) drew attention to an empirical correlation between the superionic transition temperature and the Frenkel pair formation energy of fluorite crystals:

$E/kT_c = 20$
where E is the Frenkel pair formation energy, k is the Boltzmann constant, and T_c is the superionic transition temperature.

MARKOWNIKOFF'S RULE

In the case of an asymmetrical olefin, where hydrogen halides can be added in two ways, the halogen attaches to the C atom which carries the smaller number of H atoms.

This is illustrated by Figure 39. Note that the rule (Markownikoff, 1870, 1875) may not apply when the reaction occurs in the presence of a

$$CH_3CH{=}CH_2 + HBr \rightarrow CH_3\underset{\displaystyle Br}{\underset{|}{C}}HCH_3$$

Figure 39. Illustrating Markownikoff's rule. The halogen attaches itself to the C atom which carries the smaller number of H atoms.

peroxide catalyst. Thus, vinyl bromide yields 1,1-dibromomethane under normal conditions but, in the presence of a peroxide catalyst, it yields the isomer, 1,2-dibromomethane.

MATTHIAS' RULES

On the basis of a careful study of the superconducting transition temperatures of a large number of metals and alloys, Matthias (1957) proposed the following:

1. The transition temperature of an element depends in a regular manner upon its position in the periodic table.
2. In alloys of two transition metals or of two non-transition metals, the superconducting transition temperature varies smoothly in going from one end-member to the other.

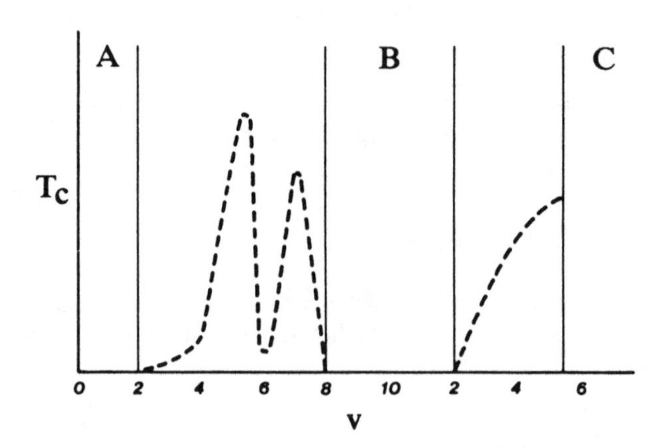

Figure 40. Qualitative variation of the superconducting transition temperature as a function of the mean valency. Region A: no superconductors, region B: no superconductors, region C: non-metallic. See Matthias' rule.

3. In elements with closed d-shells, the superconducting transition temperature increases monotonically from zero, when the valency is 2, to a maximum when the valency is 5.
4. In transition metals, the superconducting transition temperature also has a maximum when the valency is 7, has a minimum when the valency is 6, and is zero when the valency is 8.

Here, the valency referred to is the maximum valency. That is, the number of electrons outside of a closed shell. Rule 2 may not work when one or both of the components is magnetic. Rule 3 cannot be extended beyond a valency of 5 because the elements involved are non-metallic (Figure 40). This ruling out of non-metallics is somewhat ironic since, at the time of going to press, the promise of "room-temperature" superconductivity (very high transition temperature) is being offered by complex oxides.

MATTHIESSEN'S RULE

If the concentration of impurity atoms in a metal is small, the resistivity due to the scattering of electron waves by impurity atoms is independent of temperature.

This resistivity is in addition to that caused by the thermal motion of the periodic lattice of the metal. The latter resistivity is a function of temperature.

MAXWELL'S RULES

Several rules concerning electric currents and magnetic fields are attributed to Maxwell. One is that:

Every part of an electric circuit is acted upon by a force which tends to move it in such a direction as to enclose the maximum amount of magnetic flux.

Another one is the "corkscrew" rule:

The sense of the circular lines of magnetic force around a wire conductor is the same as that in which a corkscrew would have to be turned in order to drive it in the direction of the current.

Of course, it is assumed that the corkscrew is right-handed. A rule whose attribution to Maxwell is less certain concerns the magnetic polarity of a long current-carrying coil of wire:

Looking end-on at the coil of wire, draw two arrowheads which indicate the direction of the current. If one can draw an "S" between the arrowheads, that end of the coil is the magnetic south pole. If one can draw an "N" between the arrow heads, that end of the coil is the magnetic north pole.

This is best illustrated by a diagram (Figure 41). A rather tongue-in-cheek rule which has also been attributed to Maxwell is to be found under Richardson's rules. See also the swimmer rule and Fleming's rules.

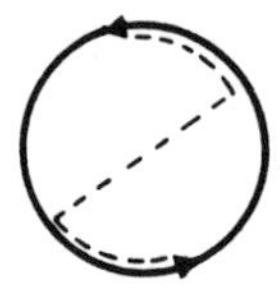
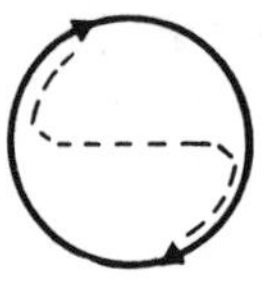

Figure 41. A Rule of Thumb for Determining the Magnetic Polarity of a Current-Carrying Coil. (See Maxwell's rules.)

McADAM'S RULE*

On the basis of the limited amount of available data concerning the elastic constants of porous materials, it was suggested (McAdam, 1951) that:

$$E = p^{3.4}$$

where E is the Young's modulus and p is the relative density.

MEISLICH RULES*

These permit one to guess at a plausible molecular orbital structure for an organic compound on the basis of the electronic structure (Meislich, 1963):

1. The hybrid orbital number, H, is defined to be equal to the sum of the coordination number and the number of unshared pairs of electrons.
2. The number of p-orbitals is equal to 4 - H.
3. The number of pi-bonds is equal to the number of pairs of electrons, minus H.
4. When it is structurally feasible, the pi-bonds interact to give extended pi-bonds.
5. It is best to use a maximum number of p-orbitals for bonding.

6. If the key atom has an unshared pair of electrons and is attached to a pi-bonded atom, hybridization occurs so as to permit the pair of electrons to enter a p-orbital and extend the pi-bond. For each pair of delocalized electrons, H is reduced by unity. Such delocalization induces a positive increase in the formal charge on the key atom and an increase in the formal negative charge elsewhere in the pi-bond.
7. Extended p-orbital overlap is marked:
   if an aromatic system results,
   if a symmetrical molecular orbital structure results,
   if the delocalization of electrons is from a weakly electronegative atom into a pi-bond with a strongly electronegative atom at the other end of the bond, or at an ortho or para position of a benzene-type ring.
8. Extended p-orbital overlap is minimal:
   if cross-overlapping with an orbital of an electron-rich atom is present and charge separation results,
   if delocalization is from a strongly electronegative element into a pi-bonded system with weakly electronegative elements.
9. Extended p-orbital overlap does not occur if adjacent like charges arise, or if a formal charge of more than + 1 or − 1 on the same atom results.

MEYER-NELDEL RULE

This describes a general relationship between the exponent and multiplying constant of the equations describing various activated processes. That is:

$$\log[P] = AQ + B$$

where P is a pre-exponential factor, Q is an activation energy, and A and B are constants.

This rule is sometimes also known as the compensation law or the theta effect.

MEYERHOFFER RULE

It is a general observation that, in a conglomerate system, the racemate is more soluble than are the constituent enantiomers. The present rule (Meyerhoffer, 1904) states that:

A conglomerate has a solubility which is equal to the sum of the solubilities of the corresponding enantiomers.

This rule was based on an analogy with vapor pressure.

MICHAEL'S RULE

This rule (Michael, 1899) applies to addition and elimination reactions in organic chemistry. The concept involves the comparison of two simultaneous competing reactions, having differing reaction velocities, which lead to two addition products that differ in structure, "handedness," or configuration. It also considers the elimination reactions of the two compounds which lead back to the same original product. The rule states that:

If either or both of two addition compounds can be formed by an addition reaction, then there is a direct relationship between the relative amounts of the two addition products formed and the velocities with which they can be reconverted into the initial substance.

That is, they are directly proportional to each other. Michael considered it to be a law rather than a rule. However, the limits of validity of the above depend upon the relationship between the activation energies and the action constants.

MID-BANDGAP RULE

The photoluminescence behavior of chalcogenide materials, of various types and with a wide range of compositions, has been found to be described by the rule:

$$E_P = E_g/2$$

where E_P is the photoluminescence peak position and E_g is the band edge value.

MOCK-GÜNTHERODT RULE*

Brillouin scattering studies (Mock & Güntherodt, 1984) of a series of rare-earth intermetallic compounds of the form, RCu_2Si_2, indicated that:

$$B = Q/V$$

where B is the bulk elastic modulus, Q is the valence of the rare earth, and V is the volume of the unit cell. It is likely that the same rule holds true for other series of intermetallic compounds.

MOGRO-CAMPERO RULE*

The period of time up to the electromigration failure of thin-film conductors depends in a complicated manner upon factors such as the current density, conductor geometry, film material, substrate type, temperature, grain size distribution, and method of deposition. It was suggested (Mogro-Campero, 1982) that:

Other factors being equal, the metal which is most susceptible to failure is that with the lowest melting point.

MOLLWO-IVEY RULE

This is a very general relationship (see Berezin's rules) and relates various transition energies to the crystal structure dimensions according to:

$$E = Ad^{-n}$$

where E is a transition energy, d is the nearest-neighbor atomic spacing, and A and n are constants.

The effectiveness of the rule implies that the details of an ionic atomic structure are essentially irrelevant and Madelung-type effects predominate.

MONTGOMERY'S RULE*

It was suggested (Montgomery, 1911) that:

The boiling point of an organic isomer is proportional to the density of the liquid at the boiling point.

MOOSER-PEARSON RULE*

This rule (Mooser & Pearson, 1956) provides an important link between those properties which are related to the energy-band structure, and the Pauling rules (qv) of chemical bonding:

In a non-metallic compound, any excess valence electrons which are not used in cation-anion bonding must be localized in non-bonding orbitals of the cations or the anions.

The rule can be used, for instance, to estimate whether a given material is likely to be a semiconductor. A graph (Mooser-Pearson plot) of average principal quantum number versus electronegativity difference also serves to divide the structures of compounds into well-defined groups. Unfortunately, there seems to be no general rule for drawing the lines of demarcation between them and consequently this plot has limited predictive value. See Pamplin's rules.

MOSS RULE

The two most interesting optical properties of a semiconductor are the absorption edge or optical energy gap, and the refractive index. As a result, attempts have been made to find a general relationship between these parameters. One such rule (Moss, 1950) for estimating one quantity on the basis of the other is:

$$n^4E_g = \text{constant}$$

where n is the refractive index, E_g is the optical energy gap, and the constant is equal to 95eV.

See also the Ravindra-Srivastava rule.

MOTT'S RULE*

Hardness is usually expressed in terms of the Vickers or Knoop hardness number. The two numbers do not differ greatly. They depend upon such factors as the anisotropy of the indented surface, the load, and the indentation time. In older publications, or in specialist applications, one finds hardness expressed in terms of the more qualitative Mohs scale. A convenient rule (Mott, 1956) for converting from one to the other is:

$$H_K \text{ or } H_V = 3M^3$$

where H_K is the Knoop hardness, H_V is the Vickers hardness, and M is the Mohs hardness.

This holds true for M-values of up to 8. See also Sangwal's rules.

MOWERY'S RULE*

At the time of its suggestion (Mowery, 1969), there was still a considerable lack of clarity concerning the reason for optical activity and also a lack of criteria for predicting the possible existence of optical isomers. He therefore devised the rule that:

Optical activity is to be expected when there is no rotation-reflection or alternating axis of symmetry in a molecule, or in any of the conformations which are produced by bond vibration or rotation; including ring alternation.

This was not intended to be an absolute criterion, but only an easier method than the comparison of mirror images. In fact, the above rule fails in the case of compounds such as (+)-menthyl - (−)-menthyl - 2,6,2′,6′-tetranitro - 4,4′-diphenate, since a molecule of this substance has no alternating axis of symmetry in any of its conformations but is, by rotation of the central diphenate grouping, identical with its mirror image. The compound is therefore optically inactive.

MUKHERJEE'S RULE*

By assuming that the vacancy formation energy is proportional to the bond energy in face-centered cubic metals, and that the Young's modulus is proportional to the bonding energy, it can be deduced (Mukherjee, 1964) that:

$E_f/Y = \text{constant}$

where E_f is the vacancy formation energy and Y is the Young's modulus.

This correlation between vacancy formation energy and modulus is suitable only for metals which have similar Poisson's ratios and atomic volumes, and which have the same form of interatomic potential. See the Reynolds-Couchman rule.

MYERS' RULES*

It has been pointed out (Myers, 1981) that the application of a few simple principles is all that is required in order to decide the coordination number and geometry of the vast majority of complexes of metal cations:

1. Ions having octets in the outermost shell, or having a filled d^{10} configuration outermost, are spherical.
2. Ions with ns^2 in the outer shell, while expected to be spherical, in fact join with 3 ligands and the unshared pair of electrons occupies the tetrahedral fourth position.

3. Metal cations which are not spherical are usually consistent in coordination number and geometry.
4. All cations not described by the first 3 rules can be expected to have a coordination number of 6; with the ligands being arranged octahedrally.

Typical ions of the type mentioned in rule 1 are: Ag^+, Au^+, and Cu^+. Typical examples of the type mentioned in rule 2 are the pyramidal $SnCl_3^-$ and $PbCl_3^-$ ions. Typical ions in the case of rule 3 are Ag^{2+}, Cu^{2+}, Ni^{2+}, Pd^{2+}, and Pt^{2+}. These usually have a coordination number of 4; with the ligands arranged in a square-planar arrangement. The Ni^{2+} ion is particularly variable, being square planar in $Ni(CN)_4^{2-}$ but tetrahedral in $NiCl_4^{2-}$ and sometimes octahedral as in $Ni(NH_3)_6^{2+}$.

NABARRO'S RULE*

When considering the interaction between two defects, it is of interest to know whether the configurational force is accompanied by a mechanical force. It was concluded (Nabarro, 1988) that:

> A mechanical force is present in the interaction between two elastic defects if, and only if, the motion of the defects under the action of their configurational force requires the motion of matter in the same direction.

NACHTRIEB'S RULE*

This is one of a number of rules which relate the activation energy for diffusion, E, to other thermodynamic parameters:

$$E/L = \text{constant}$$

where L is the latent heat of fusion. See also the Sherby-Simnad rule.

NAKAMURA'S RULES*

A survey of published data reveals (Nakamura, 1981) that the known Debye temperatures of the alkali halides obey a linear function of the form:

$$T_D{}^2 m r^3 = \text{constant}$$

where T_D is the Debye temperature, m is the average mass, and r is the nearest-neighbor distance. On the other hand, the known Debye temperatures of isostructural covalent substances, especially Si, Ge, Sn (grey), and C (diamond) obey a linear function of the form:

$$T_D^2 m r^4 = \text{constant}$$

NATANSON'S RULE

When a plane-polarized ray of light traverses an optically active medium, dextrorotation results if the right circularly polarized component is transmitted faster than the left one, and vice versa. The sign of the elliptical vibration which is produced by an optically active absorbing medium is the same as that of the circular component which is least absorbed. The present rule states that:

On the long-wavelength side of an optically active absorption band, the sign of the elliptical vibration is the same as the sign of the rotation. On the short-wavelength side, the ellipse is of opposite sign to the rotation.

See also Bruhat's rule.

NEGITA'S RULE*

In order to prepare higher-performance silicon nitride ceramics, it is necessary to find suitable sintering aids, and to understand the factors which cause such an agent to be effective. The present rule suggests that:

Metal oxides can be advantageously added to silicon nitride if the radius of the metal ion is less than 0.1nm and the electronegativity of the metal ion is less than 1.5.

Unlike most of the other rules in this book, the present one (Negita, 1985) considers a single material. However, it is a very practical and useful application of a rule which probably reflects an underlying correlation of the Hume-Rothery (qv) or Mooser-Pearson (qv) type. As such, it indicates that helpful short-cuts can be suggested by such parameter plots, as well as suggesting the basic mechanisms which are involved.

NERNST-THOMPSON RULE

In a material with a low dielectric constant, the electrostatic attraction between the cations and anions of a dissolved electrolyte will be large, and vice versa.

From this, it can be deduced that solvents with a high dielectric constant will favor dissociation whereas those with a low constant will have a small dissociative effect.

NEUMANN-KOPP RULE

The heat capacity per gram-atom of a solid phase is a weighted sum of the heat capacities of the constituents of the phase.

This rule can be expected to apply only to compounds in which the bonding and crystal structure are similar to those of the constituent elements.

NICOLET-LAU RULE*

On the basis of experiment, and literature data, it is concluded (Liu et al, 1983) that:

The ion mixing of multilayered samples will lead to the formation of an amorphous binary alloy when the two constituent metals have different crystal structures.

The relative atomic sizes and electronegativities are considered irrelevant.

NIETZKI'S RULE

The absorption of a dye is shifted towards the red end of the spectrum with increasing molecular weight of the substituted groups.

A rule which unfortunately has many exceptions.

NITROGEN RULE

An organic molecule with an even molecular weight must contain an even number of N atoms, and one with an odd molecular weight must contain an odd number of N atoms.

This rule (Beynon, 1960) is useful when deducing the empirical formula of a nitrogenous substance from mass spectra data. It depends upon the fortuitous fact that the most abundant stable isotope of the majority of elements of odd valency has an odd mass number while the corresponding isotope of those of even valency has even mass. Nitrogen is a common exception. See also the rule of 13 and Harkin's rule.

NORDHEIM'S RULE

The residual resistivity of a binary alloy is given by:

$$\text{resistivity} = cx(1 - x)$$

where x and $(1 - x)$ are the mole fractions of the two constituents, and c is a proportionality constant.

Typical exceptions to the rule are alloys of transition metals with noble metals.

OCTANT RULE

This rule was originally proposed as a method for predicting the optical rotatory behavior of substituted cyclohexanone derivatives, but has since been extended to other classes of organic compound. The basic ideal is to divide the molecule into 8 sectors, using 3 mutually perpendicular planes. In the case of cyclohexanone, the first plane contains the carbonyl O atom, the carbonyl C atom, and the 2 adjacent C atoms. The second plane contains the carbonyl O atom, the carbonyl C atom, and the opposite C atom. The third plane is perpendicular to the first 2 planes and passes through the carbonyl double bond. Having divided the molecule in this way, the rule states that:

> Looking from the direction of the carbonyl double bond, substituents in the lower-left and upper-right of the nearest block of 4 octants have a negative effect upon the optical rotation; those in the other two octants have a positive effect; and these effects are reversed for the equivalent octants in the more distant block.

Further points to remember are that the contributions of various groups in the same octant are additive; that atoms which are closer to the carbonyl bond have a greater effect; and that atoms which lie in one of the planes have essentially no effect. In certain compounds, and for certain substituent groups, the optical effects are reversed. These are described by various anti-octant rules.

OCTET RULE

In the case of all of the inert gases, except He, the outermost shell of the atom contains 8 electrons; regardless of whether the inner shells are completely filled or not. On the grounds that the inert gases are extremely unreactive, it has been suggested that an outer group of 8 electrons is an especially stable arrangement, leading to the rule that:

When atoms combine, they do so in such a way as to lead, as far as possible, to the formation of complete electron octets.

The rule has exceptions, such as SF_6 and PCl_5, where the outer shells of S and P appear to have 12 or 10 electrons, respectively.

ONE-THIRD RULE

See the 1/3 rule.

OPTICAL SUPERPOSITION RULES

This concept was originally introduced by Van't Hoff (1898) and suggests that the partial rotation which is contributed by a given asymmetric C atom to the total rotation of a molecule containing several such atoms is independent of the configuration of the other atoms. It can be summarized by the rules that:

1. In a molecule which contains several asymmetric C atoms, each of them behaves as if the remainder of the molecule is inactive.
2. The optical effects of various asymmetric C atoms in the same molecule can be added algebraically.

In principle, the application of these rules is limited to stereoisomers which differ only in the configurations of a series of asymmetric C atoms. The validity of the rule is severely limited by the phenomenon of "vicinal action." This means that, if asymmetric C atoms are close to each other, there may be interactions which change the contribution to the total rotation. A more strict, but also a more generally valid, version of the first rule is that:

Like functional groups in like surroundings make like contributions to the optical rotation.

The detailed rules of optical rotation have been greatly extended by Hudson (qv).

OSENBACH-BITLER-STUBICAN RULES*

It was shown (Osenbach et al, 1981) that the relationships between the chemical and tracer diffusion coefficients for aliovalent ions in an ionic lattice are:

$D = 2D^*$ for a dimer migration mechanism

$D = 3D^*$ for a trimer migration mechanism

where D is the chemical diffusion coefficient and D^* is the tracer diffusion coefficient.

It is assumed that the concentration of impurity ions is relatively low, but the above rules are valid regardless of the charge on the aliovalent or lattice ions.

OSTWALD'S RULE

This rule (Ostwald, 1897) states that:

If several forms of a substance can be produced in a chemical reaction, then the most labile form is produced first. Subsequently, successively more stable forms develop out of this.

This is a re-statement of an earlier (step-wise) rule due to Gay-Lussac.

OTSUKA-KOSUKA-CHANG RULE*

This rule is based upon a study of the diffusivity of oxygen in liquid metals (Otsuka et al, 1984) and states that:

The ratio of O diffusivity to self-diffusivity in liquid metals is related to the enthalpy of formation, per mole of O, of the respective oxide at 298K.

OWEN'S RULE*

Laboratory power supplies which are based upon alternating current mains usually depend upon the use of a large capacitor to reduce ripple. By storing each half-cycle of the supply, the capacitor maintains a current reservoir upon which the load can draw. Ripple in the capacitor depends upon the amount of current which is drawn from the capacitor by the load. The calculation of the ripple in the output of a supply is usually a tedious process and involves the use of design handbooks and the plotting of graphs. However, the peak-to-peak ripple voltage can easily be calculated (Owen, 1968) by means of a simple formula. That is:

> The peak-to-peak ripple voltage is equal to 7(I/C), where I is the load current in mA, and C is the filter capacitance in microfarads.

It is assumed in the above case that the supply frequency is 60Hz. For a supply frequency of 50Hz, the multiplying constant becomes equal to 8.7, while for a frequency of 400Hz the constant becomes equal to unity.

PAIRED ELECTRON RULE

This rule (Condon, 1954) was an early attempt to rationalize changes in the valency and oxidation potential of the elements as a function of position in the periodic table. It is stated in the form:

> The lability (instability) of paired electrons decreases with increase in atomic number within a group of the periodic table.

The period 2 (first-row) elements are an exception to this rule.

PAK-INAL RULES*

A study of the fracture behavior of grain boundaries in $L1_2$-type intermetallic compounds (Pak & Inal, 1987) suggested that:

1. Grain boundaries are ductile in Kurnakov compounds having an order-disorder transition below their melting point, and a wide solid-solution field.
2. Grain boundaries are brittle in Berthollide compounds having no order-disorder transition below their melting point.

A typical Kurnakov compound is Ni_3Fe and a typical Berthollide compound is Ni_3Si. The ductile grain boundary behavior in the first case is attributed to the existence of disordered grain boundaries which are intrinsically ductile and are inherited when formed in disordered states. The difference in behavior is critically important with regard to the likely strengthening effect of third elements upon Ni_3Al; a very promising high-temperature material.

PAMPLIN'S RULES

A method was suggested (Pamplin, 1964) for predicting new semiconducting compounds and alloys on the basis of mathematical expressions which were deduced from the rules of valency and from structural analogies. This led to the establishment of rules for probable semiconduction in various structures:

1. In the case of binary tetrahedral compounds ($A_{1-x}B_x$), x must be equal to (4 – a)/(b – a), where a and b are the group numbers of elements A and B, respectively.

Since x has to be positive, either or a or b must be greater than 4. This reflects the observation that the A and B elements in such a structure come from opposite sides of the Zintl boundary.

2. In the case of ternary tetrahedral compounds ($A_{1-x-y}B_xC_y$), x and y must satisfy the equation,

$$x(b - a) + y(c - a) = 4 - a$$

where a, b, and c are the group numbers of elements A, B, and C, respectively.

This permits a wide range of theoretically possible compounds, and predicts that there should be a solid solution range for valid values of a, b, and c.

PANETH'S ADSORPTION RULE

A solid material strongly adsorbs radio-isotopes which give insoluble or slightly soluble components with the electronegative (acid radical) component of the adsorbant.

This rule has many exceptions, but was improved by Hahn. See Hahn-Paneth rule.

PARTIAL ROTATIONAL SHIFT RULE

This rule states that:

An optically active alpha-amino acid which incorporates more than one asymmetric center has the L-configuration if, upon adding an acid to its aqueous solution, the partial molar rotation of the alpha-asymmetric center becomes more positive. If a negative direction of rotational shift is adopted by the alpha-asymmetric center, the amino has the D-configuration.

It is related to the Clough (qv) and Clough-Lutz-Jirgensons (qv) rules.

PASCAL'S RULE

This rule (Angus & Hill, 1943) states simply that:

In the case of diamagnetic organic compounds, the molar magnetic susceptibility is an additive property.

Common exceptions to the rule are the halogenated products of methane.

PATTERSON-BRODE RULES

It was suggested (Patterson & Brode, 1943) that an alpha-amino acid possesses an L-configuration if it obeys one of the following rules:

1. The optical rotatory dispersion is normal and positive, with the value of l_0 being greater than 205nm.
2. The optical rotatory dispersion is normal and negative, with the value of l_0 being less than 140nm.
3. The optical rotatory dispersion is anomalous, with the sign of rotation changing from negative to positive with decreasing wavelength.

Here, l_0 is the wavelength at which the graph, of reciprocal optical rotation versus the square of the wavelength, crosses the wavelength axis. An additional one, known as the modified Patterson-Brode rule states that:

4. An l_0 value of 199nm is the dividing line between positively rotating L-amino acids (above this value) and negatively rotating L-amino acids (below this value). The D-enantiomorphs have the same l_0 value but an opposite sense of rotation.

PAULING'S RULES

A series of rules which can be used to predict the structure of ionic compounds:

1. The coordination number of a metal ion, A^{n+}, increases with the radius ratio, r_A/r_B.
2. Whenever possible, charges are neutralized locally.
3. The presence of shared edges and shared faces between coordination polyhedra decreases the stability of a structure.

This rule has the corollary that:

3a. In a crystal containing various types of cation, those with a large charge and small coordination number tend not to share polyhedron elements.

Here, A is usually a metal and B is a non-metal. In rule 1, the relationship between coordination number and ionic size is complicated by factors such as the absence of alternative structures, and by the degree of polarizability of the ions. In crystals that contain complex ions, there is a reasonable correlation between the coordination numbers of cations and their sizes. The implementation of rule 2 requires the use of the electrostatic bond strength, z/n, where z is the charge on the cation and n is the number of surrounding anions. Rule 3 is based upon the fact that such sharing brings the cations of a structure closer together. This effect is large for cations with a large charge and small coordination number. Rules 3 and 3a should be applied only to complex compounds since, in simple ionic crystals, geometrical constraints over-ride them.

PFEIFFER-CHRISTELEIT RULE

It was noted that the optical rotatory behaviors of complex salts belonging to the same configurational family were markedly similar; not only with respect to the overall pattern of the dispersion curves but also with regard to the range of wavelengths within which the maximum rotation occurred. This led to the rule (Pfeiffer & Christeleit, 1937) that:

In going from lower to higher wavelengths, aqueous solutions of the copper complexes of L-amino acids become progressively more positive, reach a maximum, and then become more negative at higher wavelengths. The dispersion curves for the D-isomers are exactly the opposite to those for the enantiomorphs.

Although this rule does not necessarily apply to the total molar rotation of amino acid/Cu complexes with more than one center of symmetry, it becomes entirely applicable when it is applied only to the rotatory dispersion which is exhibited by the alpha-asymmetric center of such amino acids.

PICTET-TROUTON RULE

See Trouton's rule.

PILLING-BEDWORTH RULE

Oxide/metal volume ratios which are very different to unity imply poor oxidation resistance.

The ratio mentioned in this rule (Pilling & Bedworth, 1923) has to be carefully defined, since a simple ratio of the oxide densities is insufficient. It is defined as:

volume of oxide/volume of metal = Md/amD

where M is the molecular weight of an oxide with the formula, Me_aO_b, and density, D, and m is the atomic weight of metal with density, d. The number of atoms of metal per molecule of oxide is given by "a."

An examination of metals which exhibit (unfavorable) linear oxidation kinetics indeed reveals that these metals are exceptional with regard to their oxide/metal volume ratios (Table 15). That is, they are either less than unity, or are much higher. The assumption made is that, in the first case, the oxide is unable to form a coherent surface film, due to the low volume ratio. In the other case, the oxide is non-protective because internal stresses cause it to spall off. The metals in the first group include K, Ba, and Mg. Those in the second group include U, Nb, Ta,

Table 15
Pilling-Bedworth Ratios

Oxide	Ratio
BaO	0.67
GeO_2	1.23
K_2O	0.45
MgO	0.81
MoO_3	3.30
Nb_2O_5	2.68
Ta_2O_5	2.50
U_3O_8	2.77
WO_3	3.35

Mo, and W, but Ge is an exception. The rule is probably less significant for scales that grow by the outward migration of matter, but more important for scales in which diffusion of matter occurs from the surface towards the metal/scale interface.

PORLEZZA'S RULE*

This states that:

$T_c = 0.96(T_m + T_b)$

where T_c is the critical temperature, T_m is the melting point, and T_b is the boiling point.

The author (Porlezza, 1923) based his conclusion upon the law of corresponding states. See also Prudhomme's rule, the Lorenz rule, and the Do-Yen-Chen rule.

PRANDTL-GLAUERT RULE

The pressure coefficient on a thin airfoil in a subsonic flow with Mach number, M, is the same as the pressure coefficient in incompressible flow past a similar airfoil whose thickness is multiplied by the factor, $(1 - M^2)^{1/2}$.

A rule which arises quite simply (Houghton & Boswell, 1969) from the form of the differential equation describing the flow.

PRELOG RULE

This (Prelog, 1964) is similar to the Cram rule (qv) and states that:

A reagent will approach a ketone carbonyl group from the side with the smaller attached group.

In applying this rule, the arrangement of the groups in the starting material must first be specified, since there may be several bonds around which the molecule can rotate.

PRESTON'S RULE

This states that:

In the anomalous Zeeman effect, lines of the same series exhibit the same pattern.

PRICE-HAMMETT RULE

It was shown that the entropy of activation involved in the rates of formation and in the hydrolysis of semicarbazones is the dominant factor and bears no relationship to polar effects. This led to the formulation of the rule (Remick, 1949) that:

> In a solvent of high dielectric constant, and for a reaction having a highly polar transition state, it is expected that the more complex (entropy-containing) molecule will lose more entropy in the formation of the transition state, That is, its activation entropy will be more negative.

The rule applies, for instance, to the reaction of benzaldehyde with acetone and methyl ethyl ketone.

PRUDHOMME'S RULE

$$T_c = T_m + T_b$$

where T_c is the critical temperature, T_m is the melting point, and T_b is the boiling point.

This so-called "rule of three temperatures" (Prudhomme, 1920) was originally criticized as being entirely fortuitous. However, it remains a good general approximation. See also Porlezza's rule, the Lorenz rule, and the Do-Yen-Chen rule.

PUGH'S RULE*

The prediction of the deformation behavior of crystals requires a knowledge of their ductility as well as their fracture strength. It was proposed (Pugh, 1954) that the extent of the plastic range of a pure metal could be deduced from the ratio of the elastic bulk modulus, K, to the shear modulus, G. This gives the rule that:

> High K/G implies ductility; low K/G implies brittleness

No distinct boundary can be set for practical purposes because of all of the factors (such as strain rate, temperature, and grain size) which can affect ductility. However, the relative order of ductility in the hexagonal close-packed metals (those metals most likely to exhibit brittleness) is predicted quite accurately (Table 16). In particular, the notably malleable metal, Tl, and the notoriously brittle metal, Be, are very clearly in their correct places. If one were to hazard a guess at where to put a "ductile/brittle" barrier, a value of 2 would be reasonable. See also the Rice-Thomson rule.

Table 16
Pugh's Rule as Applied to Hexagonal Close-Packed Metals

Metal	K/G	Ductility
Tl	6.52	good
Zr	2.65	good
Ti	2.47	good
Co	2.31	fair
Cd	2.21	fair
Re	2.05	fair
Mg	2.03	fair
Hf	1.95	good
Zn	1.73	poor
Y	1.63	fair
Ru	1.63	poor
Be	0.74	poor

PULFRICH RULE

Measurements of the refractive index are often used to estimate various properties of liquids, such as the composition. During the preparation of mixtures, non-ideal behavior may complicate the prediction of refractive indices. It was found (Pulfrich, 1889) that:

> The proportional change in the complementary refractive index is equal to a constant multiplied by the change in volume upon mixing, where the complementary index is found by subtracting unity from the refractive index.

RADIUS RATIO RULES

These concern the prediction of the crystal structure which is to be expected of an ionic compound. They can be summarized as follows:

1. Since cations are usually smaller than anions, the crystal structure tends to be determined by the number of anions which it is possible to pack around the cation.
2. When the cation is very small, it is possible to pack only two anions around it; if anion-cation contact is to be maintained.
3. As the ratio of cation size to anion size increases, it becomes possible (at a critical value of 0.155) to place three anions around the cation and thus increase the Coulomb attractive energy.
4. At a critical radius ratio of 0.414, octahedral coordination becomes possible.
5. At a critical radius ratio of 0.732, eightfold cubic coordination becomes possible.

These radius ratios furnish an invaluable first approximation to the stereochemistry of ionic compounds, and a large number of structures can be interpreted in terms of them. However, the radius ratio is only one of a series of criteria which determine the structure and should not be used in isolation.

RAMSAY-YOUNG RULE

This states that:

$$(T_a/T_b) - (T_a'/T_b') = \text{constant} \times (T_a - T_a')$$

where T_a, T_b, T_a', and T_b' are the temperatures at which two substances, a and b, both have one of two different vapor pressures, P and P'.

RANK SIZE RULE

Statistics courses for scientists tend only to treat scientifically important distributions: normal, Poisson, etc. Attention is not usually paid to topics such as the Pareto curve, which describes the upper tail of a highly skewed distribution. This curve is merely an exponential one, with a negative exponent. The rank size rule, which is a special case of the Pareto curve, can be useful when one is confronted by a sample which is too few in members, has too skewed a distribution to be treated using normal methods, and in which individual members of the population can be identified. For example, the separation of immiscible liquids with decreasing temperature may produce very large globules, plus others whose sizes trail off rapidly into the microscopic range. In such a case, it would be worthwhile to use the present rule:

$$P_r = P_1/r$$

where P_r is the size of an identifiable member of the population, P_1 is the size of the largest member of the population, and r is the rank of the member in the population. That is, the second-largest member is likely to have a size which is half of that of the largest member, the third-largest member is likely to have a size which is one third of that of the largest member, and so on.

See also Benford's rule.

RAVINDRA-SRIVASTAVA RULE*

Two important optical properties of a semiconductor are the absorption edge or optical energy gap, and the refractive index. Consequently, attempts have been made to find a general relationship between these parameters. A convenient rule (Ravindra & Srivastava, 1979) for estimating one of these parameters on the basis of the other is:

$$n^4E_g = \text{constant}$$

where n is the refractive index, E_g is the optical energy gap, and the constant is equal to 108eV.

Another relationship, $n = 4.084 - 0.62E_g$, was proposed by the same authors. However, this is not a rule which is easily remembered. See also the Moss rule.

REACTING BOND RULE

This rule (Swain & Thornton, 1962) predicts that:

In a series of reactions involving different N compounds, the nearest (C-N) reacting bond will become longer as the group which is attached to the N atom becomes more electron-donating. At the same time, the more remote (C-O) bond will become shorter.

That is, the transition state will become more like a reactant and the carbonyl O will become less basic. According to the solvation rule (qv), the proton should then be more remote from this O atom, but closer to the anion.

REDDY'S RULE*

$$C = B - A + 1$$

where C is the number of rings, B is the number of ring bonds, and A is the number of ring atoms in a cyclic molecule.

This equation (Reddy, 1987), which is a statement of Euler's formula, can be used to check a tentative systematic name. Thus, in the case of cubane (A = 8, B = 12), the fact that C is equal to 5 means that it is a pentacyclic system, whereas adamantane (A = 10, B = 12, C = 3) is tricyclic (Figure 42). See also Eckroth's rule and Trahanovsky's rule.

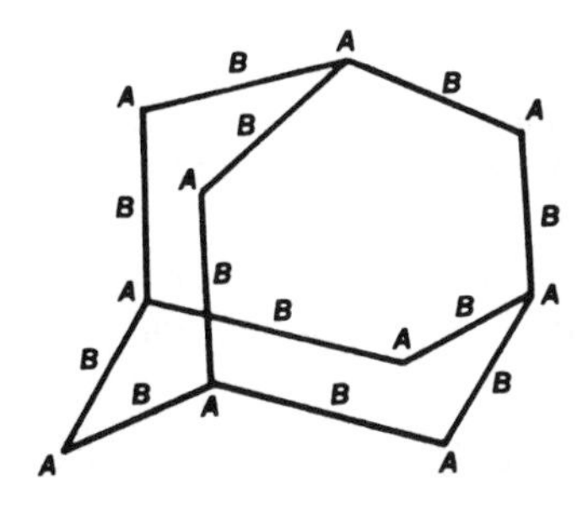

Figure 42. Using Reddy's rule to determine the number of rings in the cyclic molecule, adamantane. Here, the number (A) of ring atoms is 10, the number (B) of ring bonds is 12, and the number (C) of rings is therefore 3; making it tricyclic. This example is fairly trivial. The power of Reddy's rule is more evident when applied to very complicated molecules.

REYNOLDS-COUCHMAN RULE*

According to the Mukherjee rule (qv), the vacancy formation energy is proportional to the Young's modulus. However, this correlation holds only for metals which have a similar Poisson's ratio and atomic volume. A more elaborate rule (Reynolds & Couchman, 1974), which takes account of these parameters, is:

$$E_f/VY(1 - 2u) = \text{constant}$$

where E_f is the vacancy formation energy, V is the atomic volume, Y is the Young's modulus, and u is the Poisson ratio.

REYNOLDS-COUCHMAN-KARASZ RULE*

On the basis of the Couchman (qv) rule and one of the Gorecki rules (qv), it was suggested (Reynolds et al, 1976) that surface energy can be estimated from:

$$sr^2/T_m = \text{constant}$$

where s is the surface energy, r is the nearest-neighbor separation, and T_m is the melting point.

RICCI'S RULE

This is simply an expression which relates the ionization constant of a material to its structural characteristics, such that:

$$-\log[K] = 8 - 9m + 4n$$

where K is the ionization constant, m is the formal charge on the central atom, and n is the number of non-hydrogenated O atoms.

This rule can be used in "reverse" in order to deduce the structure of an inorganic compound from the ionization constant.

RICE-THOMSON RULE*

It was suggested (Rice & Thomson, 1974) that the ductility of crystals is related to the width of the dislocation core. This is because dislocations with larger cores tend to be more mobile and facilitate plastic flow at a crack tip, thus decreasing the tendency to brittleness. This gives the rule that:

Brittleness is directly proportional to Gd/E

where G is the shear modulus, d is the interatomic spacing, and E is the surface energy.

Attentive readers will have noticed that this is the same as the first of Kuczynski's rules (qv). However, the latter author made no comment about brittleness. The present rule seems to order the hexagonal metals quite accurately (Table 17). The positions of the extreme cases of malleable Zr and fragile Be are clearly correct. The intervening metals are less well ordered by this rule than they are by Pugh's rule (qv).

Table 17
Use of the Rice-Thomson Rule to Predict the Ductility of Hexagonal Metals

Metal	Gd/E	Ductility
Zr	6.90	good
Ti	7.31	good
Mg	8.10	fair
Hf	9.16	good
Co	9.27	fair
Cd	10.24	fair
Y	11.59	fair
Zn	11.73	poor
Re	15.73	fair
Ru	19.45	poor
Be	26.12	poor

RICHARDS' RULE

This states that:

H_m/T_m = constant

where H_m is the molar heat of fusion and T_m is the melting point in degrees Kelvin.

The constant is approximately equal to 2. This rule was one of the two which were mentioned to the compiler during university courses. However, it was wrongly described as being Trouton's rule (qv). The latter applies to vaporization and is more widely obeyed.

RICHARDSON'S RULE

The factor by which the activity coefficient of a metallic solution is lowered by a given atomic percentage of a solute is proportional to the difference between the heats of formation of the solute's oxide and of the host's oxide.

This rule (Richardson, 1950) is reported to be well-obeyed and can be used to estimate the effect of a solute upon the activity of a third element in a metallic solution.

RICHARDSON'S RULES*

Like me, you probably often have thought that it is a great pity, when solving field problems, to derive the general solution, force it to fit the boundary conditions, and then possibly have to plot the results. All that work is required to obtain a set of contours of equal concentration (or temperature, or pressure, or magnetic field strength), which is much as one might have guessed. Is there not an easier way to obtain the same result? This is a problem that must have exercised L. F. Richardson for, in 1908, he gave two very simple rules for solving the Laplace equation in orthogonal coordinates:

Given that the contours satisfy the conditions imposed at the boundaries, the Laplace equation is automatically satisfied by a freehand-drawn network of lines in the interior if:
1. The corners of the network are orthogonal, and
2. The network, when sufficiently finely divided, consists of squares.

All that one needs in order to produce reasonably accurate results is a pencil and eraser, or a blackboard. The method is very simple and effective when applied to problems involving Dirichlet or Neumann boundary conditions. It becomes more tedious when the boundary conditions are of Robin type and involve a functional relationship between the absolute boundary values and their gradients. The rules can also be modified so as to treat the Laplace equation in other coordinates, or to treat other related equations. However, although the rules have only to be slightly modified in these cases, the technique becomes so awkward in practice that the use of more modern methods is preferable. Even so, it is a fruitful way of whiling away a long train or aeroplane journey and the accuracy is certainly no less than that of the method which was suggested by Maxwell. His advice (Elementary Treatise on Electricity and Magnetism) was tentatively to alter known solutions of the Laplace equation by drawing diagrams on paper and then to select the least improbable of them. An impressive result, obtained using the present rule, is shown in Figure 43. One of the best overviews of the present method (flux-plotting), as used in its hey-day, is provided by Poritsky (1938).

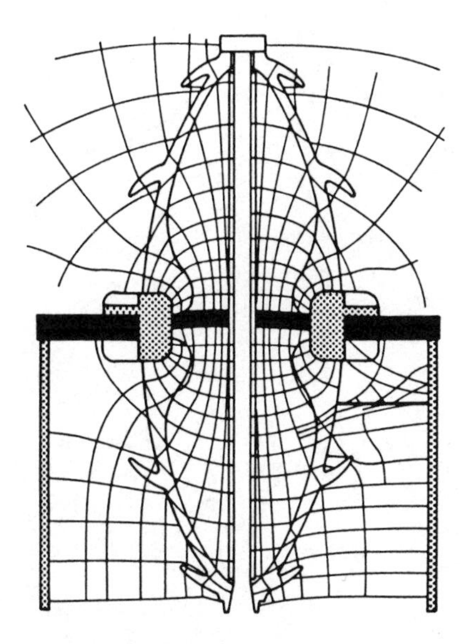

Figure 43. A spectacular example of the use of the flux-plotting method (see Richardson's rules) for rotational symmetry. Here (Kuhlmann, 1914), a transformer bushing design has been analyzed in spite of the complications caused by the fact that 3 dielectrics (air, oil, porcelain) are involved. Note the refraction effects at their interfaces. The left-hand side is the field for a low oil level, and the right-hand side is the field for a high oil level (thin black horizontal line).

ROUX-VIGNES RULE

A rule (Roux & Vignes, 1970) which appears to be of very wide applicability to metallic alloys, and which predicts qualitative trends in solute diffusivity, is that:

The elements which diffuse more rapidly than the solvent are those which, when dissolved in the pure solvent, lower its melting point. Likewise, those elements which diffuse more slowly are the ones which increase the melting point.

RUNGE'S RULE

In the complex Zeeman effect, the frequency shift of a line in a magnetic field is either equal to a Balmer line, or to a/b times it, where a and b are small integers.

SANGWAL'S RULES*

Hardness is expressed in many ways in the literature. One of the most common scales used is the Vickers hardness. Two less-used and related scales are those of Mohs and scratch hardness. A convenient rule (Sangwal, 1982) for converting from one to the other is:

$$M = H_s^{1/4}$$

where M is the Mohs hardness and H_s is the scratch hardness.

Often, the reason for measuring the hardness is not to determine this value itself but rather to estimate some property which is related to it. For instance, the Vickers hardness of ionic crystals is related to the surface energy by the rule:

$$H_V/E^2 = \text{constant}$$

where H_V is the Vickers hardness and E is the surface energy.

The Vickers and Mohs hardness values can be related using Mott's rule (qv).

SAVILLE'S RULES

It is observed that many nucleophilic and electrophilic substitution reactions in organic chemistry are not just simple single displacements but in fact involve simultaneous electrophilic and nucleophilic attack on a substrate. This gives rise to a 4-center reaction in which the attacking electrophile and nucleophile mutually assist each other. The attacking species can be independent or coupled. Also, attack can occur at an isolated or conjugated multiple bond as well as at a single bond. This gives rise to addition instead of displacement. The present rules are "merely" the SHAB rules (qv) for inorganic reactions, as applied to the selection of the optimum combination of attacking electrophile and nucleophile for a given organic reaction.

SAVITSKY'S RULES*

The ability to visualize clearly a complicated network of points in three dimensions is a gift which is given to very few people. Sections through relatively simple crystal structures such as the face-centered cubic one are easily pictured; provided that the Miller indices of the sectioning plane are of low order. The atomic arrangements on planes with higher Miller indices are much more difficult to imagine, as are the arrangements in complicated crystal structures (such as Mn) even when sectioned using planes with low Miller indices. However, the visualization of these arrangements is often essential in fields such as metallurgy, where one may wish to know the type and distribution of atoms on slip planes, or surface reaction chemistry, where the surface composition or geometry of an exposed crystal may determine the kinetics. The present rules (Savitsky, 1960) help enormously in problems which involve crystal surfaces. In order to determine the make-up of an arbitrary crystalline plane, all one has to do is:

1. Divide the unit cell into n + 1 equally spaced planes, parallel to the upper surface, in such a way that all of the atoms (assumed to be points) lie on one of the planes (Figure 44).
2. Spread out the planes, in order, in such a way that the vertical axes of the contiguous planes are aligned (Figure 45).
3. Displace successive planes, by distances equal to A, 2A, 3A, etc., in the direction of the horizontal axis (Figure 46).
4. With the origin at the top left of the first plane, draw a line which passes through the origin and through a point on the lower edge of the top plane which is a distance, B, from the lower left (Figure 46).
5. Deduce which atoms lie on the arbitrary section by extending the above line through all of the n + 1 planes. Lines which are parallel to the original one should also be considered.
6. Optionally, the desired sectioning plane is transferred back onto the original unit cell by taking advantage of the distance-preserving character of the rules.

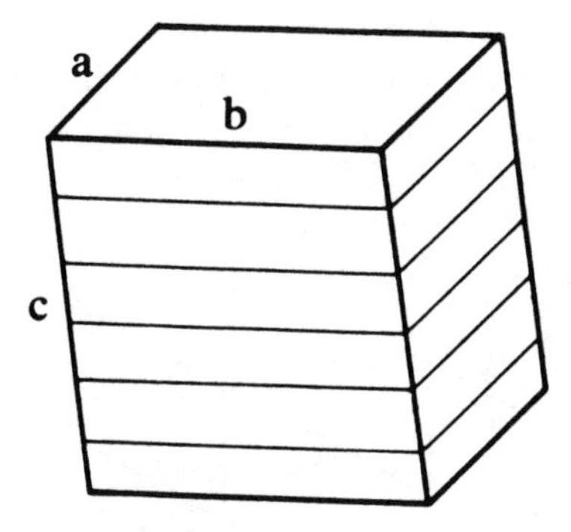

Figure 44. Use of Savitsky's rules. A (trigonal) unit cell is sectioned using planes which are parallel to its upper surface.

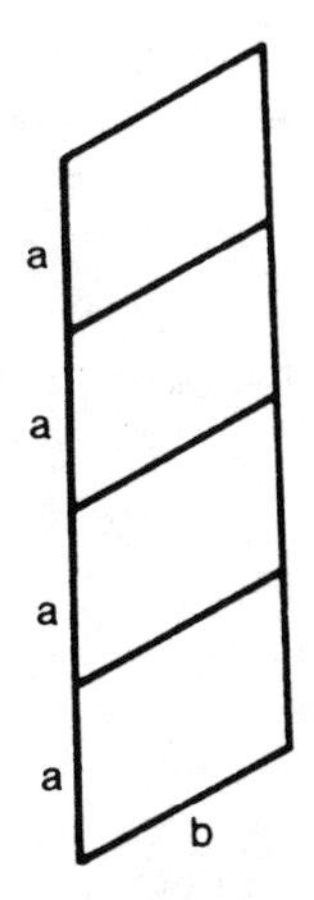

Figure 45. Use of Savitsky's rules. The a-axes of the contiguous slices are aligned.

The only disadvantage in the use of the rules is the slight difficulty of remembering the definitions of A and B:

$$A = b(nh + l)/nk$$

$$B = bh/k$$

where (hkl) are the Miller indices (arranged in monotonically increasing order) of the sectioning plane, and b is a unit cell parameter. The rules are completely general. For instance, it has been assumed in Figures 44 to 46 that the unit cell is of trigonal type. In Figure 47, the rules have been used to determine the nature of the (123) planes of CaF_2.

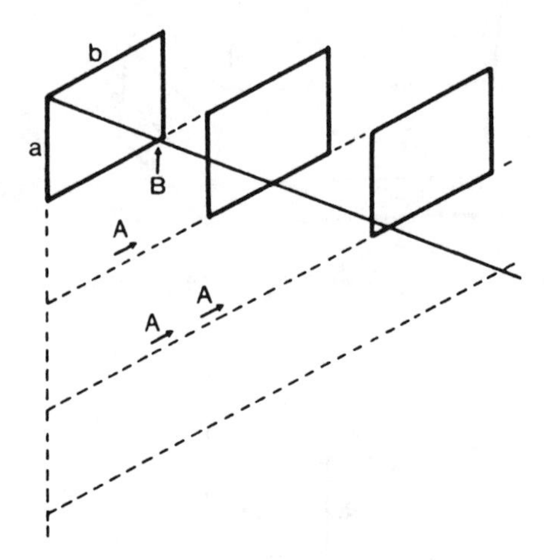

Figure 46. Use of Savitsky's rules. The slices are displaced by A, 2A, etc., and a line is drawn which passes through the origin and through a point whose position is determined by B. See text for the definitions of A and B.

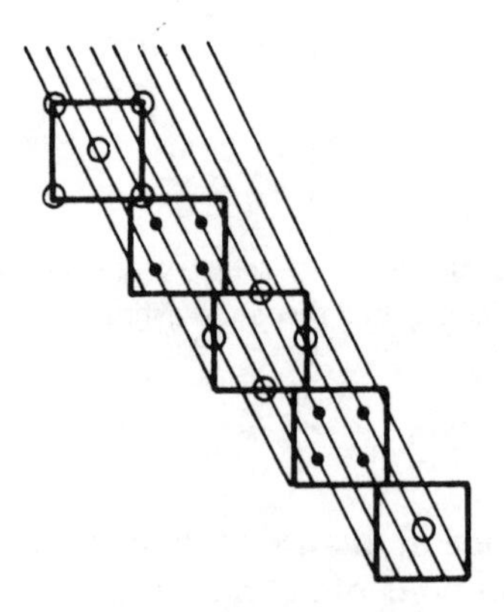

Figure 47. Use of Savitsky's rules. Application of the rules to a study of the (123) planes of CaF_2.

SAYTZEFF'S RULE

The simplest statement of this rule (Saytzeff, 1875) is:

In the dehydration of alcohols, H is eliminated preferentially from the adjacent C atom which is poorer in H.

This is illustrated by Figure 48. Every chemistry teacher who mentions this rule seems impelled to follow it with an apposite quote from the Bible ("... but from him that hath not shall be taken away even that which he hath"). This rule means that the tertiarily bound H of an R_3CH structure is more reactive than the H of a methylene group, R_2CH_2, and the latter is more reactive than the H of the primary C atom of a methyl group, RCH_3. The rule is also applicable to the elimination of a hydrogen halide from an alkyl halide. A more general statement of the rule is that:

The olefin which is produced in the greatest abundance is that in which the maximum number of alkyl groups is attached to the double bond.

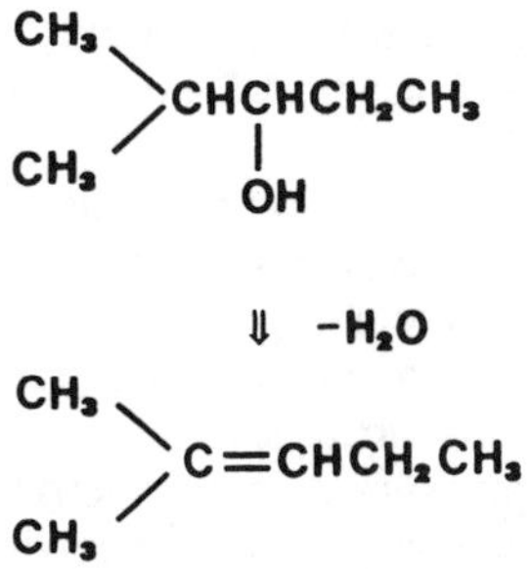

$$(CH_3)_2CHCH(OH)CH_2CH_3 \xrightarrow{-H_2O} (CH_3)_2C{=}CHCH_2CH_3$$

Figure 48. Illustrating Saytzeff's rule. During the dehydration of an alcohol, the second H atom is taken preferentially from the adjacent C atom which is already poorer in H.

SCHÄFER'S RULES*

These general principles (Schäfer, 1964) can be applied to chemical transport processes involving heterogeneous equilibria. They are either self-evident or can be understood with the aid of a simple diagram (Figure 49). Thus:

1. A reaction can support transport only when no solid substance is present on one side of the equilibrium equation.
2. A reaction with an extreme equilibrium position gives no measurable solid material transport. Therefore, in the selection of transport sys-

tems, one can immediately impose the fundamental condition that the equilibrium condition must not be extreme.
3. The sign of the reaction enthalpy determines the transport direction. Exothermic reactions give transport from a lower temperature to a higher temperature, and endothermic reactions give transport from a higher temperature to a lower temperature. When the reaction enthalpy is zero, the pressure difference is also zero and no transport can occur.
4. Unless the reaction entropy is zero, there is a value of the reaction enthalpy which gives a maximum transport effect.
5. When the reaction entropy is very far from zero, transport is possible only when the reaction entropy and enthalpy have the same sign. The transport becomes large when the logarithm of the equilibrium constant is approximately equal to zero.
6. If the reaction entropy is small then, depending upon the sign of the reaction enthalpy, transport can take place from a lower temperature to a higher temperature, or vice versa.
7. For a reaction with a sufficiently large positive change in entropy, one can expect appreciable transport of a solid substance only from a higher temperature to a lower temperature. A reaction with a negative entropy change can lead only to transport from a lower temperature to a higher temperature.
8. The maximum tendency to transport increases with increasing absolute value of the entropy change or when the enthalpy change varies correspondingly; both leading to a larger difference in absolute pressure.
9. Changing the equilibrium constant by altering the temperature can considerably affect the transport process.
10. An increase in the pressure difference can be obtained by choosing the gas composition so that the reaction equilibrium shifts to a less extreme position. When the reaction involves a change in the number of moles present, manipulation of the total pressure can have the same effect.

Rule 6 is especially true when no change in the number of moles in the gas phase occurs. For small values of the reaction enthalpy, the maximum transport effect will correspond to small but finite values of the logarithm of the equilibrium constant. When the reaction entropy, reaction enthalpy, and equilibrium constant are all equal to zero, the degree of transport is also zero; consistent with rule 3. The temperature effect in rule 9 is marked because of the steepness of the curves in Figure 49. The rules have been proved many times to hold in the selection of transport reactions. Because the entropy change of a reaction can be estimated from the number of participating gas molecules, the values of the reaction enthalpy and the temperature which are required in order to ensure appreciable transport are known. See also the Van't Hoff rules.

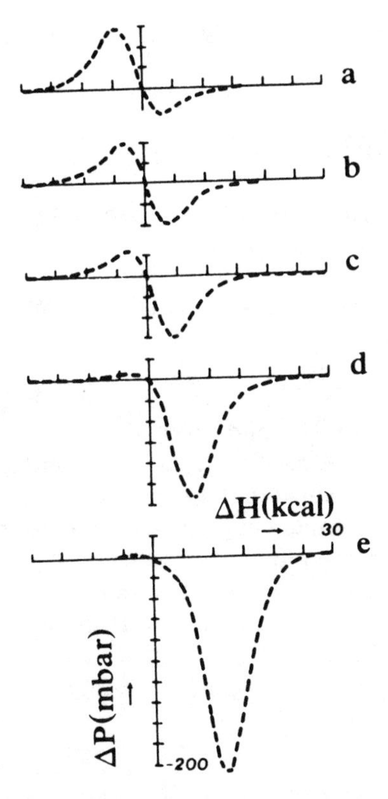

Figure 49. Curves summarizing the chemical transport behavior of a solid substance, via the reaction: A(solid) + B(gas) = C(gas), as a function of the reaction enthalpy and entropy. Vertical scale is pressure difference of component B at the lower and higher chamber temperatures, and reflects the degree of transport which occurs. Horizontal scale is reaction enthalpy. The reaction entropy increases from a to e. These curves explain many aspects of chemical reaction transport. See Schäfer's rules.

SCHMIDT'S RULES

The influence of the double bond upon the strength of single bonds in organic compounds is summarized (Schmidt, 1934) by the rules that:

1. A double bond weakens, above all, the next alternate simple bond, while the directly adjacent single bond is strengthened.
2. For still further removed single bonds, there is an alternate weakening and strengthening so that the 3rd, 5th, (2n + 1)th bonds are stronger while the 2nd, 4th, (2n)th bonds are weaker than normal.
3. The weakening of the second or next alternate bond is reflected by the fact that fission takes place at this point during thermal decomposition.

4. Fission of a chain never takes place at a directly adjacent single bond during cracking.

SCHREINEMAKERS' RULE

This rule (Schreinemakers, 1915) is essentially the same as Wilson's first rule (qv), and states simply that:

> Phase boundaries, when produced, must extend into fields with a higher number of phases.

However, it has been shown more recently (Hillert, 1985) that the rule holds true even under conditions where it has no clear theoretical justification, and can be applied to many-component phase diagrams. For instance, it is well known that the topology of a two-dimensional section through a multi-component phase diagram can be deduced from the phases which are observed in alloys (Gupta et al., 1986). Unfortunately, the so-called "zero phase fraction" boundaries which are found in this way do not behave properly at their intersections (Figure 50a). The use of Schreinemakers' rule (Hillert, 1988) solves this difficulty and produces a more realistic diagram (Figure 50b).

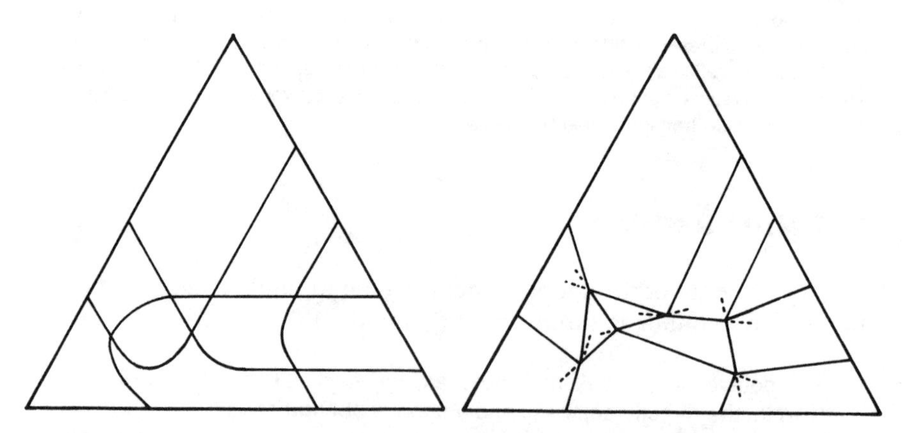

Figure 50. The use of Schreinemakers' rule to improve the construction of phase diagrams. In Figure 50a (after Gupta et al., 1986), a first approximation to a two-dimensional section of a multi-component phase diagram has been made. In Figure 50b (after Hillert, 1988), the diagram has been improved by ensuring that Schreinemakers' rule applies at every intersection.

SCHULZE-HARDY RULE

Only the ions of opposite charge to that on the colloidal particle are important, and their coagulating power increases sharply with increasing valency of the ion.

This rule was established by a number of workers but is named after only two of them (Schulze, 1882/3; Hardy, 1900). The minimum concentrations of trivalent, divalent, and univalent opposite-charge ions required to produce rapid coagulation (precipitation of essentially all of the colloid in a few seconds) are approximately in the ratio, 1:10:500 (Table 18). The valency and the nature of the ions with the same charge as that of the colloidal particles has very little effect upon the coagulating power.

Table 18
Illustrating the Schulze-Hardy Rule
(Minimum concentrations required for the rapid coagulation of colloidal sulphur)

Salt	Concentration (mol/l)
HCl	6
LiCl	0.913
NaCl	0.153
KNO_3	0.022
KCl	0.021
RbCl	0.016
$MgSO_4$	0.0093
CsCl	0.009
$AlCl_3$	0.0044
$CaCl_2$	0.0041
$Sr(NO_3)_2$	0.0025
$BaCl_2$	0.0021

SHAB RULES

This (SHAB) is one of several acronyms which have been used in the literature when referring to the hard/soft acid/base concept of Pearson. The rules state that:

1. With regard to equilibrium, hard acids prefer to associate with hard bases and soft acids with soft bases.
2. With regard to kinetics, hard acids react readily with hard bases and soft acids react readily with soft bases.

Although these rules are easily remembered, it is sometimes much more difficult to remember whether a given acid or base is supposed to be soft or hard. In addition, there are borderline cases between hardness and softness for both acids and bases, as well as amphoteric substances which can act as both an acid or a base.

SHER-CHEN-SPICER RULE*

A theoretical study of the dislocation energy and hardness of semiconductors (Sher et al., 1985) suggested that:

If a tetrahedrally bonded semiconductor is alloyed with one having a shorter bond length, the dislocation density should be decreased.

This conclusion was based upon theoretical and experimental evidence that the dislocation energy in these materials is inversely proportional to the bond length raised to a power of between 3 and 9. Also, the hardness is inversely proportional to the bond length raised to a power of between 5 and 11. The value of the exponent depends upon the relative degrees of ionicity and covalency of the bonding.

SHERBY-SIMNAD RULE*

If diffusion occurs via a vacancy mechanism, and if the diffusivity at the melting point is a constant, according to the Van Liempt rule (qv), it follows (Sherby & Simnad, 1961) that:

$$E/RT_m = \text{constant}$$

where E is the activation energy for diffusion, R is the gas constant, and T_m is the melting point.

SHEWMON'S RULES

Some of the most-quoted rules of thumb are those given by Shewmon (1963) for the estimation of diffusion data:

$E/T_m = 36cal/K$

$E/L = 16.5$

$D_m = 10^{-8}cm^2/s$

where E is the activation energy for diffusion, L is the latent heat of fusion, and D_m is the diffusivity at the melting point. However, these rules are essentially the same as those of Sherby & Simnad (qv), Nachtrieb (qv), and Van Liempt (qv).

SHIFT RULE

See Freudenberg's rule.

SHINSKEY'S RULES*

Many pieces of laboratory equipment, such as furnaces, incorporate devices known as controllers which keep the equipment at some status quo. The setting of the various dials on controllers can be very tedious. There are many ways of tuning controllers, and elaborate mathematical procedures have been devised in order to find the optimum settings. Shinskey (1971) suggests a simple method which is based upon a knowledge of the natural frequency of the system. He gives rules for producing a stable loop under three-mode control:

1. Adjust the re-set time to its maximum value, and the derivative time to its minimum value.
2. Reduce the proportional band until oscillation begins. The period, in minutes, of this oscillation is determined by measuring the time between successive peaks.
3. The re-set and derivative values are then both set equal to the above period, divided by 2π. The proportional band is then adjusted so as to obtain the desired degree of damping.

In the case of a two-mode controller, re-set introduces a phase lag which is not counteracted by a derivative term. The best re-set time is then equal to the natural period of oscillation, divided by 2.

SHIRLEY-HALL RULE*

The electronic interaction energy between H and substitutional impurities was estimated (Shirley & Hall, 1984) using an approximate phenomenological model. The electronic interaction was found to be attractive when the metal impurity had an electron deficit with respect to the host, and repulsive when the impurity had an electron excess. It was found that, because of the general relationship between atomic size and the number of valence electrons, the two interactions tend to occur in a complementary fashion so that:

> Impurities to the left of the host in the periodic table trap hydrogen and those to the right do not.

This rule bears some resemblance to the diagonal rule.

SHUKLA'S RULE*

It is well known that the observed shear modulus of a polycrystal must lie between the Reuss and Voigt limits. The latter limits are complicated weighted averages of the monocrystalline elastic constants. For the case of cubic crystals, Shukla (1982) proposed the rather simpler rule that:

> The maximum and minimum values of the shear modulus of a polycrystal are set by the highest and lowest values of the shear modulus of a single crystal.

It was found that Si was a notable exception to this rule.

SIDGWICK-POWELL RULES

1. When there are no unshared valence electron pairs about a central atom, the coordinating atoms will take up positions which are as far apart as possible.
2. Unshared pairs of electrons will occupy bond positions.

These rules (Gould, 1955) summarize a great deal of important information about the directional character of chemical bonds. The first rule is analogous to the problem of minimizing the total energy of a small number of point charges on the surface of a sphere. The effect of the second rule is best shown by a diagram (Figure 51).

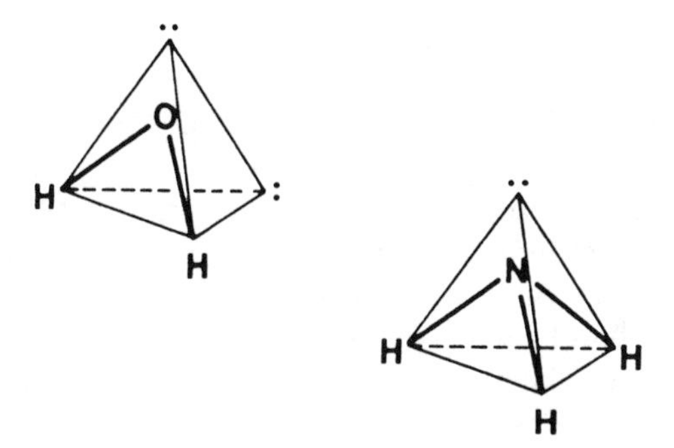

Figure 51. Sidgwick-Powell Rules. Unshared pairs of electrons will occupy bond positions.

SIGAREV-GALIULIN RULES*

It was suggested (Sigarev & Galiulin, 1985) that there are two general rules which a material should satisfy in order to be a superionic conductor:

1. The possible paths of the ions in the crystal are equivalent to each other.
2. The paths of the ions are such that their movement breaks the symmetry of the crystal to a minimal degree.

The first rule implies some form of tetrahedral arrangement, while the second rule implies that the ions must travel along symmetry elements of the crystal.

SLIDE RULES*

Without doubt, a major source of irritation at scientific conferences is the parlous state of many of the slides which are shown. In particular, they are often illegible when seen from anywhere but the front row. The best way of avoiding this fault in one's own slides is to check them by using the rule that:

If a slide is not readable when held at arm's length, it will not be readable when projected from the back of a conference hall.

Any slide which does not obey this rule should be rejected. Judging by the confusion which occurs at most conferences, there also appears to be a large number of speakers who still believe that slides suffer lateral inversion during projection. Therefore, while reading the slide at arm's length it is well to remember that it only has to be rotated by 180° in its own plane before projection, rather than being rotated by 180° about the vertical or horizontal axis. Finally, having correctly ordered and oriented the slides chosen for a lecture in their carrying case, it is a good idea to draw a straight line diagonally across the tops of the slides using a wax crayon. That is, draw a line from the left-hand corner of the first slide to the right-hand corner of the last slide. Should they accidentally become jumbled, it is then easy to re-order them quickly, without reference to the lecture notes.

SMELTZER'S RULE*

It has been found (Smeltzer et al., 1970) that, for metals:

The erosion rate of a metal depends upon the melting point.

See also the rules of Ascarelli, Brauer & Kriegel, Finnie-Wolak-Kabil, Hutchings, Khruschov, and Vijh.

SMITH RULE

Combining rules are often used (Stwalley, 1971) in order to estimate the interaction potential, P_{AB}, of asymmetric pairs of atoms or molecules when those for the symmetric pairs (P_{AA}, P_{BB}) are known. The most commonly used rules are the arithmetic mean combining rule (qv) and the geometric mean combining rule (qv). However, these rules have disadvantages when the difference in the sizes of A and B is large. This is because the interactions in all three systems are assumed to occur at equal values of R. As a result, Smith (1972) proposed a combining rule, for the repulsive interaction, which took account of this size difference. Attention was restricted to the closed-shell repulsive interactions because accurate theoretical combining methods exist for the long-range forces (Kramer & Herschbach, 1970). The rule is based upon the assumption that the electron densities of the two atoms or ions are distorted as they approach each other, and that the repulsive energy is the sum of the distortion energies of the two atoms or ions. According to this rule:

The asymmetric pair interaction potential between two atoms or ions is given by:

$$P_{AB}(R) = [P_{AA}(2r_A) + P_{BB}(2r_B)]/2$$

where $r_A + r_B = R$, and $P_{AA}(R)$ is the A-A interaction potential at separation, R, etc.

Here, r_A is the distance from nucleus A to an imaginary boundary between the two species, and is found from the equation, $dP_{AB}/dr_A = 0$. The above expression reduces to the arithmetic rule when $r_A = R/2$. What the Smith rule does is to incorporate, into the arithmetic mean, the fact that the two atoms or ions in the AB pair are different. Lee and Kim (qv) have modified this rule.

S_N1 RULE

This well-known reaction mechanism, a first-order unimolecular nucleophilic substitution, has a stereochemical reaction course which is summed up by the rule (Ingold, 1953) that:

> The S_N1 mechanism, proceeding through a carbonium ion, usually involves racemization with an excess of inversion unless a configuration-holding group is present. In the latter case, the configuration is predominantly preserved.

A suitable configuration-holding group would be the alpha-carboxylate ion.

S_N2 RULE

This well-known reaction mechanism, a second-order bimolecular nucleophilic substitution, has a stereochemical reaction course which is summed up by the rule (Ingold, 1953) that:

> Substitution by the S_N2 mechanism involves an inversion of the configuration; regardless of the constitutional details.

This is also facetiously known as the "umbrella in a gale" rule because the chemical bonds are flipped over like the ribs of an umbrella in a strong wind.

SOLIDUS RULE

See the Roux-Vignes rule.

SOLVATION RULE

This rule (Swain et al, 1965) predicts that:

A proton which is being transferred, in an organic reaction, from one O or N atom to another should lie in an entirely stable potential at the transition state and not form reacting bonds or give rise to primary H isotope effects. It should lie closer to the more basic atom (O or N) at the transition state, but be increasingly remote when substituent changes make this atom less basic.

SOMORJAI-SZALKOWSKI RULES*

Low-energy electron diffraction studies have revealed that chemisorption leads to mainly ordered structures on monocrystalline surfaces. The type of surface structure which forms depends markedly upon the symmetry of the substrate, the size and chemical nature of the adsorbed gas molecule, and sometimes upon the surface concentration of the adsorbate. The available experimental data were analyzed (Somorjai & Szalkowski, 1971) in order to determine the rules which governed the formation of ordered surface structures:

1. Adsorbed atoms or molecules tend to form surface structures which are characterized by having the smallest unit cell which is permitted by the molecular dimensions, and the adsorbate-adsorbate and adsorbate-substrate interactions. Close-packed arrangements are preferred.
2. Adsorbed atoms or molecules form ordered structures which have the same rotational symmetry as that of the substrate plane.
3. Adsorbed atoms or molecules, at monolayer thickness, tend to form ordered structures which are characterized by having unit cell vectors which are closely related to the substrate unit cell vectors. As a result, the surface structure bears a greater similarity to the substrate structure than to the structure of the bulk condensate.

These are known as the "rule of close packing," the "rule of rotational symmetry," and the "rule of similar unit cell vectors," respectively. Difficulties arise when the first rule is used because there are certain molecules which exhibit more than one binding state on a given surface; such as CO on the (100) surfaces of some metals. A notable exception to the third rule occurs when the adsorbed layer exhibits partial ionization. Mutual repulsion may then lead to adsorption as a disordered open structure. This is the case for adsorbed Na on W surfaces.

STALKER-MORRAL RULES*

In ternary and higher diffusion systems, extrema can occur in the concentration profiles. The extrema are important because they play a large role in the occurrence of zero-flux planes and can affect the experimental determination of diffusion coefficients. It was shown (Stalker & Morral, 1988) that:

1. n-component diffusion couples can have a maximum of 2n – 4 extrema in their concentration profiles.
2. In systems with constant diffusivity, the total number of extrema must be an even number.
3. In systems with variable diffusivity, there can be an even or an odd number of extrema.

STEVENSON'S RULE

This states (Stevenson, 1951) that:

When a molecule can dissociate via two alternative processes,

$R_1R_2 \rightarrow [R_1R_2]^+ \rightarrow R_1^+ + R_2$
or
$R_1R_2 \rightarrow [R_1R_2]^+ \rightarrow R_1 + R_2^+$

fragments of the first sort will be produced in their lowest states, or without kinetic energy, only when the ionization potential of R_1 is less than that of R_2.

This means that bond dissociation energies can be calculated only for the process which gives the ion of lower ionization potential. See also the breakdown rules.

STOHMANN'S RULE

According to this rule (Stohmann, 1889):

The variations in the dissociation constants and the heats of combustion should be the same for a series of isomeric acids.

This suggests that there is a relationship between the work of ionization and the energy content of the substances. Because there can be constitutional features which markedly change the energy content, but hardly affect the dissociation constant, there must obviously be exceptions.

STONEHAM'S RULES*

An important question in many applications is whether a given liquid metal will wet an oxide substrate. A number of criteria have been identified (Stoneham, 1983a) as being important in determining the wetting behavior. For instance, wetting is favored by a low electron density of the metal, and by a small insulator band-gap. Trace impurity effects are also expected to be pronounced whenever the impurity can react with the insulating substrate. However, the most intriguing results are the rules that:

1. A liquid metal wets a non-metal substrate only when the refractive index of the latter exceeds a critical value of about 2.2.
2. Strong metal/support interaction in a supported metal catalyst occurs only when the refractive index of the supporting non-metal exceeds a value of about 2.2.
3. Strong metal/support interaction in a catalyst occurs only if the liquid metal would wet the substrate surface.

SUNBERG'S RULE*

The number of ppm (parts per million) of a compound at a concentration of 0.001M is equal to the molecular weight.

This rule (Sunberg, 1986) means that, for instance, 0.001M HCl has a concentration of 36.5ppm. Knowing this, it is then easy to deal with other molarities.

SWALIN'S RULES*

These suggest that, in an alloy consisting of A and B atoms (where the former are larger),

1. The diffusion activation enthalpy of A atoms in a matrix of B atoms is greater than that of B atoms in a matrix of B atoms.
2. The probability of an abundance of vacancies in the neighborhood of an A atom is greater than of an abundance of vacancies in the neighborhood of a B atom.

SWIMMER RULE

The relationship between the direction of the current in a wire, and the resultant direction of deviation of a parallel magnetic needle, was expressed by Ampère in the form of a picturesque rule:

> If one imagines that one is swimming in the direction of the electric current, then the movement of the north pole of the needle is the same as that of the left arm.

No doubt due to the later proliferation of new swimming strokes and the consequent ambiguity, this rule is nowadays stated in the form:

> Hold the right hand palm-down above the wire, with the fingers pointing in the direction of the current. The north pole of the magnet will be deflected in the direction of the extended thumb.

This is sometimes referred to as the "right hand rule." Both this and the swimmer rule are obviously special cases of, and less succinct than, Fleming's rules (qv).

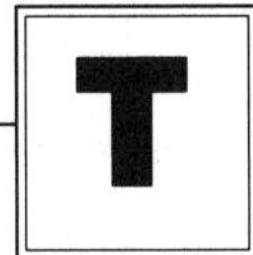

TABOR-MOORE RULE*

This rule (Tabor, 1970; Moore, 1974) relates the mechanical and elastic properties of a metallic glass, according to:

$$H = CY$$

where H is the hardness, Y is the Young's modulus, and C is a constant which is approximately equal to 3.

TAFT RULE

On the basis of a separation of the potential energy and kinetic energy effects involved in ester hydrolysis, the rule was proposed (Taft, 1953) that:

> If a group such as H or CH_3 is replaced by a group which has many more internal degrees of freedom, and if the activation process is such that these groups are compressed into positions which result in greater repulsion for the latter than for the former substituent, then the activation process will also be accompanied by a greater loss of internal motion.

The converse is not necessarily true.

TAMMANN'S RULES

An early attempt to rationalize the formation of compounds. They state that:

1. Except for the first two short periods, neighboring elements in the periodic system do not form compounds.

2. An element either forms compounds with all members of a characteristic group of the periodic system or with none.

Upon seeing these rules, exceptions will immediately spring to the mind of any metallurgist. In fact, they are among the least well-obeyed rules in this book. They are included only out of historical interest, for the sake of completeness, and as an early warning to anyone who may start generalizing along these well-worn lines. However, Tammann also suggested some more successful rules concerning diffusion. These state that:

1. Diffusion in metals becomes significant at 33% of the melting point in degrees Kelvin.
2. Diffusion in ionic materials becomes significant at 57% of the melting point in degrees Kelvin.
3. Diffusion in covalent materials becomes significant at 90% of the melting point in degrees Kelvin.

The rules may be useful in choosing annealing temperatures for phase diagram determinations or microstructural homogenization, but not for choosing mechanical deformation conditions. Even in metals, the minimum hot-working temperature is taken to be equal to 50% of the melting point in degrees Kelvin, and plastic deformation is expected only at much higher homologous temperatures in the other materials.

THOMSON'S RULE

This rule (Newton, 1963) is useful because it provides a means for calculating the approximate decomposition potential for a reaction. It states that:

$$E = \text{constant} \times (H/n)$$

where E is the decomposition potential in volts, H is the heat of reaction in calories, and n is the valence charge.

It is an approximate relationship because not all of the chemical energy is always converted into electrical energy. However, it can be used as a first-order guess when the additional data which are required in order to apply the Gibbs-Helmholtz equation are lacking.

TIWARI'S RULE*

$S_m < 6.7$ J/molK implies anomaly

It was suggested (Tiwari, 1978) that metals which have an entropy of fusion which is less than 6.7J/molK tend to exhibit an anomalous diffusion behavior. Here, the term, "anomalous," refers to the fact that they do not obey the usual correlations (see Le Claire's rules) between diffusion and melting parameters (Table 19).

Table 19
Entropy of Fusion and Diffusion Behavior of Metals
(see Tiwari's rule)

Metal	Entropy of Fusion (J/molK)	Behavior
Ag	9.16	normal
Al	11.55	normal
Au	9.39	normal
Cd	10.42	normal
Co	9.16	normal
Cu	9.62	normal
Fe	7.63	normal
K	6.94	anomalous
Li	6.61	anomalous
Mg	9.71	normal
Mo	11.26	normal
Na	7.0	anomalous
Ni	10.13	normal
Pb	7.99	normal
Pu	3.1	anomalous
U	6.07	anomalous
Zn	10.57	normal

TRAHANOVSKY'S RULE*

It was suggested (Trahanovsky, 1971) that:

In order to determine the number of rings in a bridged cyclic organic molecule, count the number of bond cleavages which are needed in order to convert the cyclic molecule into an acyclic one.

Although this rule is simple to apply to bicyclic or tricyclic systems, it becomes cumbersome in the case of larger bridged cyclic systems. See also Eckworth's rule and Reddy's rule.

TRAUBE'S RULES

The literature recognizes two rules which can be attributed to Traube, although each one is invariably quoted separately as though it were the only one. One concerns surface tension, and states that:

> In dilute solutions, the concentration at which equal lowering of the surface tension is observed decreases three-fold for each additional CH_2 group in a given homologous series.

This follows (Traube, 1891) from his observation that the difference in the surface tensions of solutions of a substance having two different concentrations is proportional to the molecular weight of the solute. The other rule which is attributed to Traube (1899) states that:

> The molar volume of an organic compound can be approximated by summing the atomic contributions.

Obviously, this rule is not very useful if one does not have a table of suitable molar volumes at hand (Table 20); especially as the rule does not mention the 12.4ml "co-volume" of the molecule, which must also be added.

Table 20
Atomic Molar Volumes for use with Traube's Rule

Atom	Type	Molar Volume(ml)
Br	-	17.7
C	-	9.9
Cl	1	13.2
H	-	3.1
I	-	21.4
N	cyano group	13.2
O	hydroxy[2]	2.3
O	carbonyl	5.5
O	ether	5.5
O	carboxyl group	5.9
S	thioether	15.5
S	thiocarbonyl	15.5

[1] *In monohalogenated compounds.*
[2] *Plus 0.4ml per second hydroxy O on the same or a nearby C atom.*

TRIANGLE RULE*

See Kier's Rule.

TROUTON'S RULE

This rule states that:

L_e/T_b = constant

where L_e is the latent heat of evaporation and T_b is the boiling point.

Since the left hand side is the definition of entropy, it can be said that the entropy of evaporation is the same for all substances. Its value is 22cal/molK. This relationship was deduced theoretically by Pictet (1876), and was found empirically by Trouton (1884). It breaks down only in the case of substances with a low boiling point, or when there is association or dissociation of the substance in the gaseous state.

TSCHUGAEFF RULE

This rule (Tschugaeff, 1898) states simply that:

The molecular optical rotation values of the members of a configurationally related homologous series move progressively in the same direction towards a limit or maximum and show little variation once the latter is reached.

On the other hand, even a small chemical change in the substituent which is directly adjacent to an asymmetric C atom affects the rotation according to the shift rule (qv).

TWO-THIRDS RULE

See the $^2/_3$ rule.

URBACH'S RULE

This empirical observation (Urbach, 1953) governs the behavior of optical absorption. It states that:

> Near to an absorption edge, the absorption coefficient depends exponentially upon the incident photon energy. The proportionality coefficient in the exponent is often close to unity.

A simple model (Bussemer, 1979) has been developed in order to explain the reason for this behavior.

VAN ARKEL RULE

This rule (Van Arkel, 1932), which is also known as the dipole rule, states simply that:

> The isomer with the higher dipole moment has the higher physical constants; regardless of the heat content.

It complements the Auwers-Skita rule (qv).

VAN DER LAAN'S RULES*

Systematic relationships have been identified (Thole & Van der Laan, 1987) between spin-orbit splitting in the valence band and the branching ratio in X-ray absorption spectra. They can be summarized by the rules:

1. In the absence of spin-orbit splitting in the valence band, the branching ratio is always statistical. Therefore, a crystal or ligand field (hybridization) can never produce a non-statistical value.
2. In the presence of spin-orbit interaction in the valence band, the branching ratio (averaged over all J levels of a given LS term) is statistical.

3a. As long as L and S are good quantum numbers then, within a given LS term, the largest branching ratio is obtained for the level, $J = L - S$, for less than half-filled shells.

3b. As long as L and S are good quantum numbers then, within a given LS term, the largest branching ratio is obtained for the level, $J = L + S$, for more than half-filled shells.

3c. For the other J levels, the branching ratio gradually decreases with J.

Rule 2 implies that the branching ratio will remain at the statistical value if the spin-orbit splitting in the valence band is small when compared to the band width or the field splitting. Rules 3a and 3b are simply the Hund's rule (qv) ground state.

VAN DYCK-COLAÏTIS-AMELINCKX RULES*

It is well known that some crystal systems can accommodate deviations from a simple composition by forming parallel twin planes or glide reflection planes which are arranged in the form of large super-cells. The distance between the twin interfaces can change as a function of composition. This mechanism is called polysynthetic twinning and is similar to the crystallographic shear mechanism for the accommodation of compositional changes. Simple rules have been developed (Van Dyck et al., 1981) which aid the rapid recognition of a microtwinned structure or texture, and permit the twin arrangement to be deduced from a diffraction pattern. The kinematic diffraction pattern of such a system often exhibits a characteristic X-shape in which the strongest satellites are located in the vicinity of the original Bragg reflections of both untwinned variants. Thus, bands of intense satellites are found in the neighborhood of the original reciprocal lattice rows. The two most intense bands make up the characteristic X-shape which passes through the origin. Figure 52 represents a typical diffraction pattern for a polysynthetically twinned crystal. The rules proposed for the analysis of such a pattern are:

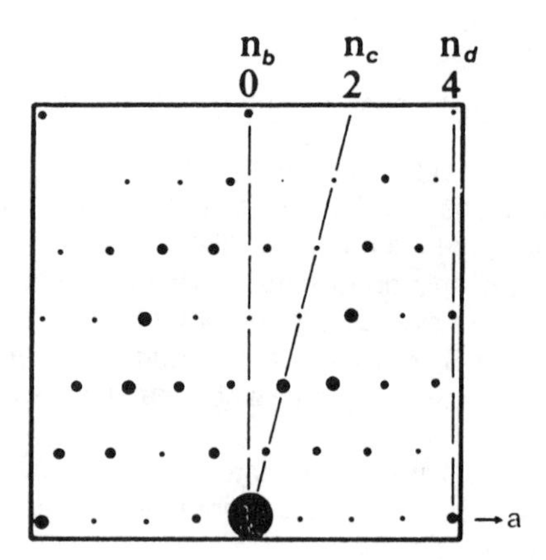

Figure 52. Schematic Electron Diffraction Pattern of an YSeF Polytype.(See the Van Dyck-Colaitis-Amelinckx Rules.)

1. Draw an axis, a, through the origin and parallel to the satellite rows.
2. Construct an axis, b, through the origin and perpendicular to a.

3. Align an axis, c, with a systematic line of satellites which passes through the origin.
4. Place an axis, d, parallel to the b-axis and passing through the first major reflection along the a-axis.
5. The d-axis also passes through the strongest satellite on another row. Call this row "m."
6. The numbers of satellites which are cut off on row m, by b, c, and d, are equal to n_b, n_c, and n_d, respectively.
7. The number of twins in the basic repeat unit of the structure is then equal to n_d, and the difference in the two types of twin is equal to n_c(mod m).

In the case shown, $n_d = 4$ and $n_c = 2$(mod 4). Thus, a possible twin arrangement would be AAAB (number of twins in pattern = 4, difference in number of twins = 2). However, because of the modular arithmetic nature of n_c, there are more possible solutions than those which are immediately apparent. Some typical solutions are shown in Table 21. Finally, the two arms which make up the X-shape of the diffraction pattern may have differing intensities due to the differing volume fractions of the two twin variants. On the other hand, the intensities are not important and the present method should also be applicable to dynamic electron diffraction patterns.

Table 21
Possible Interpretations of Diffraction Patterns
(See the Van Dyck-Colaitis-Amelinckx Rules)

n_d	n_c	Possible Solutions
4	2(mod 4)	3,1
6	0(mod 6)	3,3
6	2(mod 6)	4,2 or 1,5
10	2(mod 6)	6,4 or 9,1 or 3,7
12	2(mod 6)	7,5 or 10,2 or 4,8 or 1,11
14	0(mod 6)	7,7 or 4,10 or 1,13

VAN LIEMPT'S RULE

The present rule (Van Liempt, 1935) was one of the first to be proposed for diffusion processes, and states that:

The rate of diffusion at the melting point is a constant.

The constant was originally believed to be a universal one, but later research showed that the value was constant only for a given crystal structure. With increasing precision of the diffusion data, various subgroups and sub-sub-groups have been identified; each one having its own constant.

VAN'T HOFF RULES

The use of Le Chatelier's principle in chemistry involves several difficulties. For example, simple statements of the principle are vague, and the more precise statements in terms of intensive and extensive variables are hard to remember. One alternative is to discard the principle and apply the present rules. The Van't Hoff rules for disturbed equilibrium are:

1. For closed systems at constant pressure, and initially at equilibrium, the temperature and the amount of substance on the higher enthalpy side of the equation rise and fall together.
2. For closed systems at constant temperature, and initially at equilibrium, the pressure and the amount of substance on the lower volume side of the equation rise and fall together.

These rules are deduced (Bent, 1965) from the first and second laws of thermodynamics, and can be simplified even further to give:

1. Higher temperature favors higher enthalpy.
2. Higher pressure favors smaller volume.

When the entropy for the reaction is known, the first rule can be stated in the alternative form:

1. For closed systems at constant pressure, the higher entropy side of the equation is favored by an increase in temperature.

As an example, consider ice and water at equilibrium. If the temperature is raised at constant pressure, ice melts because water has a higher enthalpy. On the other hand, if the pressure is raised at constant temperature, then ice melts because water has a smaller volume. Another rule which is sometimes attributed to Van't Hoff (Findlay, 1951) is that:

The least solvated compound is most stable at high temperatures.

Other rules with the same name concern optical rotation. In this compilation, they are listed under their alternative name of "optical superposition rules."

VAN UITERT'S RULES*

It has been shown (Van Uitert, 1981) that, for metallic bonding,

$$T_m d^2/V = \text{constant}$$

where T_m is the melting point, d is the bond length, and V is the number of valence band electrons per atom.

It has also been shown (Van Uitert, 1979) that, for screw-like chain structures,

$$aT_m^2 = \text{constant}$$

where a is the coefficient of thermal expansion.

See also Carnelly's rule.

VAROTSOS RULES*

It has been noted (Lazaridou et al, 1985) that in halides:

1. The entropy scales with the enthalpy for the processes of Schottky formation, free cation migration, and the reorientation of cation-vacancy dipoles.

Moreover, for conduction in various halides, it is suggested (Varotsos et al, 1984) that:

2. The migration volume is proportional to the corresponding migration enthalpy.

VIJH RULE*

It has been reported (Vijh, 1975) that:

The resistance of a pure metal to sliding wear is proportional to the metal-metal bond energy.

See also the rules of Ascarelli, Brauer & Kriegel, Finnie-Wolak-Kabil, Hutchings, Khruschov, and Smeltzer.

VON AUWERS & SKITA RULE

See the Auwers-Skita rule.

VORLANDER'S RULE

This rule (Vorlander, 1919) is based upon the general concept that aromatic compounds which contain an atom of electropositive character that is directly attached to the benzene ring are all inert and meta-substituting, while those with an electronegative atom which is adjacent to the aromatic ring are reactive and are ortho- or para-substituting. More specifically, it states that:

> If the first addition to a benzene ring contains an unsaturated valency then the group is meta-orienting, and if it is unsaturated then ortho-para substitution will occur.

The rule has exceptions. For example, when the group is CH = CHCOOH (cinnamic acid) it is meta-, rather than ortho-para orienting.

WAGNER-HAUFFE RULES

One method of protecting metals against corrosion is to add elements to the base metal, so as to modify the properties of the protective film which is naturally formed and make it more effective. The amounts of solute added are usually relatively small. They depend upon the solubility in the surface compound and upon the free energies of formation. The latter determine the extent to which the second metal is preferentially oxidized. Under these conditions, the effect of additional elements can be estimated (Hauffe, 1951):

1. The oxidation rates of metals forming oxidation layers consisting of excess-metal semiconductors can be decreased by adding metals of higher valency than that of the base metal.
2. The oxidation rates of metals forming oxidation layers consisting of metal-deficient semiconductors can be decreased by adding metals of lower valency than that of the base metal.
3. The oxidation rates of metals forming oxidation layers consisting of pure cationic conductors can be decreased by adding metals of higher valency than that of the base metal.
4. The oxidation rates of metals forming oxidation layers consisting of pure anionic conductors can be decreased by adding metals of lower valency than that of the base metal.

These rules apply to an appreciable proportion of simple metallic compounds; especially oxides and sulphides.

WALDEN'S RULE

In an electrolyte, the product of the equivalent conductivity at infinite dilution and of the viscosity of the solvent is a constant.

The rule fails when the effective diameter of the ions in various solvents varies due to differing degrees of solvation.

WALLER'S RULE

According to this rule for organic materials:

The ratio of the viscosities at the freezing and boiling points is low for molecules of high symmetry and high for molecules of low symmetry.

WATSON'S RULE

This rule (Watson, 1931) states that:

$T_e/T_c = 0.283v^{0.18}$

where T_e is the temperature at which the molar concentration of the saturated vapor is equal to that of the ideal gas, T_c is the critical temperature, and v is the molar volume at the boiling point.

WEERTMAN'S RULE*

According to this rule (Weertman, 1981):

A metal is ductile if the ratio of the theoretical maximum tensile stress to the theoretical maximum shear stress is greater than 7; and is brittle otherwise.

WERNER'S RULE

This rule (Werner, 1912) for coordination compounds is analogous to Winther's rule (qv) for organic compounds and states that:

If one combines chiral complex ions, which have the same type of structure but possess different central atoms, with the same optically active substance then ions having the same configuration will give diastereometric salts which have comparable solubilities. That is, they will be uniformly higher or lower than those of the salts which are obtained using the other optical isomer.

Like Winther's rule, it can lead to false conclusions if the experimental conditions are not carefully specified. A more recent statement of the rule is that:

If two ions form isomorphous less-soluble diastereomers having the same resolving power, then they have related configurations.

WILSON'S RULES*

These were suggested (Wilson, 1944) as a handy guide to the detection of errors in phase diagrams. He was writing at a time when published diagrams for metallic systems still contained many elementary errors of construction. Since then, metallurgists have considerably improved in this respect. However, organic chemists continue to commit heinous errors, and one still sees complaints (from metallurgists) that rule 4 in particular is not often used (Pelton, 1988).

1. In eutectic (oid) and peritectic (oid) reactions, the phase boundaries must inflect towards the tie-line at the point where they meet it.
2. At a maximum or minimum in the liquidus (polymorphic) reaction, the boundaries of both phase fields must be horizontal at the point of contact.
3. The boundary of a solubility gap must be horizontal at the point of closure.
4. When a polymorphic reaction occurs in one of the pure components, the initial slopes of the phase boundaries meeting at the transition temperature differ by H_t/RT_m^2, where H_t is the latent heat of the transition, T_m is the absolute temperature, and R is the gas constant.

Rule 4 has several corollaries:

4a. When the liquidus slope is negative, the initial slope of the liquidus must be greater than H_t/RT_m^2.
4b. When the liquidus slope is positive, the initial slope of the liquidus cannot be less than H_t/RT_m^2.

It can also be used to deduce Hume-Rothery's rule (qv) for phase diagrams. See also Schreinemaker's rule for an extension of rule 1 above.

WINTHER'S RULE

On the basis of the limited data which were available at the time, it was suggested (Winther, 1895) that, in a series of structurally related compounds which are combined with the same resolving agent, the more or less soluble salt always corresponds to enantiomers of like configuration. This led to the rule that:

Two organic acids which are precipitated by the same base must have like configurations.

This rule was originally used to predict the correct relative configurations of acids such as tartaric, lactic, malic, and mandelic.

WOLF'S RULE

Two conformations, the boat and half-chair, are possible in delta-lactones. According to this rule (Wolf, 1965):

> The n-pi* Cotton effect occurs at a wavelength which is slightly lower for the boat-form than for the half-chair form, and the sign of this effect is determined by the chirality of the boat or half-chair conformation of the delta-lactone.

The original rule has since been extended to lactones in general.

WOODSUM'S RULE

The WKB approximation of quantum mechanics finds use in other disciplines (acoustics, hydrodynamics, oceanography) which involve the study of waves. One, sometimes tedious, step which must be performed when using the WKB method is the normalization of the associated wave functions. The present rule (Woodsum, 1978), expressed in quantum mechanical terms, states that:

> $G_n^2 = (dE/dn)/\pi$, where G_n is a normalization constant, E is the total energy, and n is a quantum number.

In order that this rule should apply, the denominator of the WKB wave function should vary much more slowly than does the numerator, and the contributions to the normalization integral which arise from the exponential tail of the wave function should be negligible.

WÖSTYN'S RULE

Information on the specific heats of metallurgical slags are often difficult to find. The present rule (Newton, 1963) states that:

> The specific heat of a solid slag is the weighted average of the specific heats of the constituent oxides.

The average specific heat of solid slags is about 0.25, and that of liquid slags is about 0.4. As a further guide to the properties of such slags, one can assume that the heat of fusion ranges from 40 to 80cal/kg. Glassy slags have essentially no heat of fusion. The average heat content of fused slags is 350 to 400cal/kg.

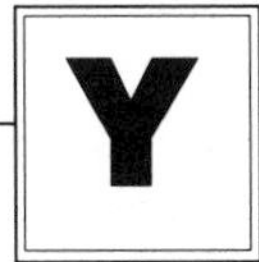

YAVARI-ROUAULT RULES*

These authors (Yavari & Rouault, 1986) have assembled a number of criteria for the prediction of metallic glass-forming compositions in metallic systems. It is expected that glass formation will be easy if:

1. There is fast interstitial diffusion of one element in the other's crystal lattice.
2. There are strong interatomic interactions.
3. There is a large atom size difference between the two constituents.
4. The composition is near to a eutectic or is of low-liquidus type.
5. There is an absence of solid solubility in the crystalline phase which is closest in composition to that of the liquid alloy.
6. There are low-symmetry intermetallic crystalline phases in the neighboring composition range and/or structures with large unit cells; indicating low-symmetry short-range order.

The first 3 factors serve to stabilize the liquid and, via the development of short-range order in the disordered phases, lower the free energy of the glassy structure towards that of the crystalline counterpart. The fourth criterion permits a more rapid quench rate from a higher-density liquid alloy. The last two rules mean that the nucleation and growth of crystalline phases is more difficult. Although there can be exceptions, the above criteria are almost always sufficient to ensure metallic glass formation.

YI'S RULE*

Writing the ground state electron configurations of the elements, according to the aufbau principle, is a standard exercise. Such configurations are also the starting point for some of the other rules in this compilation. Various methods have been proposed for remembering these configurations. The first of these was that due to Yi (1947). In this method, the various levels are arranged as in Figure 53, and one follows the arrows in order to obtain the successive electronic configurations. It has various drawbacks as a mnemonic, such as the "un-natural" method of reading the configuration. See also the Carpenter rule and the Hovland rule.

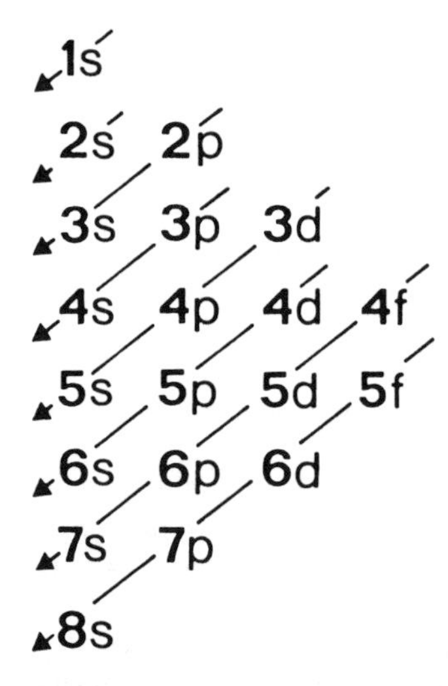

Figure 53. The Yi rule for writing the ground state electron configurations of the elements. The configurations are obtained by following the arrows.

YOSHIDA'S RULE*

On the basis of a statistical analysis of data concerning the fundamental quantities which influence surface formation in metals, it was proposed (Yoshida, 1984) that:

> The surface entropy is equal to about one tenth of the specific vaporization entropy, where the latter is defined as the vaporization entropy per unit area of monatomic layer in the bulk.

The surface entropy is not, in itself, a practically important quantity. However, it was found that the surface tension was proportional to the surface entropy, raised to the power of 1.62. The surface tension was also given by an expression of the form, $a + b(E/V^{2/3})$, where E is the vacancy formation energy and V is the atomic volume. Because of the arbitrary values of the exponents and constants involved, the two latter expressions are not proposed as rules of thumb.

ZACHARIASEN'S RULES

According to the network theory of Zachariasen (1932) and Warren (1933), the following selection patterns hold for the formation of spatial arrangements of low order. That is, for the formation of glasses from simple compounds such as SiO_2, B_2O_3, P_2O_5, GeO_2, As_2S_3, etc.:

1. An oxide or a compound tends to form a glass when its smallest building unit easily forms polyhedral building groups.
2. Any two such polyhedra must not have more than one corner in common.
3. An anion (such as O^{2-}, S^{2-}, or F^{-}) must not be linked to more than two central atoms of a polyhedron. The anions of all of the simple glasses thus form bridges between two polyhedra.
4. The number of corners of a polyhedron must be less than 6.
5. At least three corners of a polyhedron must be linked with adjoining polyhedra.

If attention is restricted to oxides, then the rules for easy glass formation can be slightly simplified to give:

a. Each O ion should be linked to no more than 2 positive ions.
b. A small number (3 or 4) of O ions should surround a positive ion.
c. Oxygen polyhedra should share corners with each other, but not edges or faces.
d. Enough polyhedra share at least 3 corners in common and make up a three-dimensional network of bonds.

If large cations are built into simple glasses which obey the above rules, then bridge failures occur in which the O which is introduced with the large cation occupies the free corner of the separated tetrahedra while the large cation fills the larger vacancy which is produced by the open lattice at this position (Figure 54). The splitting and the incorporation of large cations into the cavities of the network is assumed to take place statistically and uniformly. Furthermore, the cations can be divided into 3 groups:

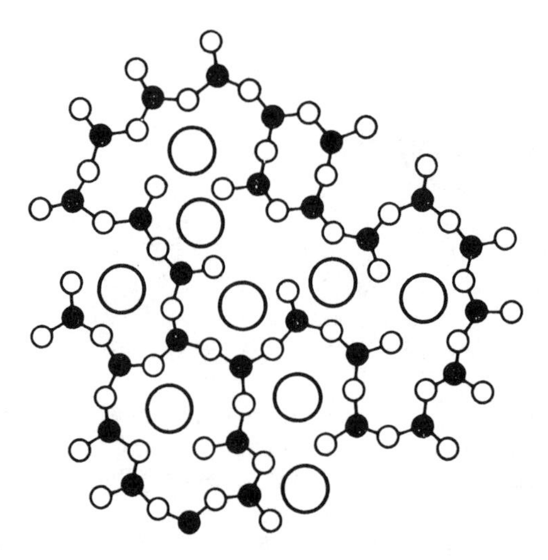

Figure 54. Illustrating Zachariasen's rules via the arrangement of ions in soda glass. Upon incorporating Na_2O into SiO_2 (quartz glass), the large Na ions (larger open circles) occupy the large cavities which result from the bridge-breaking effect of the O atoms (smaller open circles) which are also added.

1. Network formers (Si, B, P, Ge, As, Be). That is, mainly having coordination numbers which are equal to 3 or 4.
2. Network modifiers (Na, K, Ca, Ba). That is, mainly having a coordination number which is equal to 6.
3. Intermediate oxides (Al, Mg, Zn, Pb, Nb, Ta). That is, mainly having a coordination number which is between 4 and 6.

The intermediate oxides are, as their name implies, intermediate in effect between network formers and modifiers. In the case of a multi-linked glass, they may either reinforce the glass structure as tetrahedra-forming units with a coordination number of 4 or loosen the structure still more as modifiers with a coordination number of between 6 and 8.

ZEHE-RÖPKE RULE*

The formation of voids in metallic conductors, due to direct current electromigration, is an undesirable phenomenon. This rule (Zehe & Röpke, 1986) relates the formation or avoidance of voids during electromigration to the valencies of the matrix and the alloying addition. It states that:

> The formation of voids is enhanced if the difference between the valency of the matrix metal and the valency of the alloying addition is an even number. The formation of voids is suppressed when the difference between the valency of the matrix and that of the alloying addition is an odd number.

Thus, the addition of In ($5s^2 5p^1$) or Ag ($5s^1$) to Cu ($4s^1$) should lead to void formation during direct current electromigration. The addition of Sn ($5s^2 5p^2$) should not have this effect. The same is true of Si ($3s^2 3p^2$) in Al ($3s^2 3p^1$) and of O ($2s^2 2p^4$) in Ag ($5s^1$).

ZENER'S RULE

This correlation (Zener, 1951) was established for interstitial solutions and is based upon the concept that the migration free enthalpy is due to the elastic work required to strain the lattice so that the interstitial atom can jump. The rule can also be applied to self-diffusion, and states that:

$$ST_m/BE = \text{constant}$$

> where S is the entropy of diffusion, T_m is the melting point, E is the activation energy, and B is the dimensionless temperature dependence of the shear or Young's modulus.

The constant is equal to 0.55 for face-centered cubic metals and is equal to unity for body-centered cubic metals.

ZIEMANN'S RULES*

Ion beam bombardment can sometimes be used to produce amorphous metallic systems. The present author (Ziemann, 1985) has suggested a number of criteria for predicting whether a given system can be forced into an amorphous state by using ion beam techniques:

1. A chemically active implanted species tends to stabilize disorder.
2. A critical number of stabilizing implanted atoms must be exceeded in order to obtain the amorphous phase within a given volume.

3. In some cases, an amorphous phase can be produced at a given temperature but not at a higher temperature. However, if the amorphous phase is produced at the lower temperature it remains stable to above the higher temperature.
4. If a system can be produced in amorphous form by splat cooling, then ion implantation will also give the amorphous phase.

In rule 1, stabilization of disorder eventually gives rise to an amorphous phase. Inert elements, such as Ar or self-ions, produce long-range distortion but do not produce an amorphous phase. In rule 4, the reverse is not necessarily true.

ZINTL'S RULE

In salt-like ionic compounds, only elements which precede the noble gases by between 1 and 4 places are able to become negative ions.

The rule (Zintl et al., 1931) fails in the case of In and Ga alloys, but can be patched up by extending the limit to 5 places before the noble gases. Current practice seems to accept that the so-called "Zintl line" runs between the column, C-Si-Ge-Sn-Pb, on the right-hand side and the column, B-Al-Ga-In-Tl, on the left-hand side.

Numerical Rules

1/3 RULE

For many years, it has frequently been observed that:

The fracture of a wire specimen during a tensile test tends to occur at a distance of one third of the length of the specimen from either end.

Recent work (Soliman et al., 1986) has suggested that this occurs because of a complex interaction between dislocation propagation and the triaxial stress fields that are set up by the force which is exerted by the clamps on the ends of the specimen. The second most common position is at the center of the specimen. In both cases, the favored breaking point is associated with a peak in the hardness.

2/3 RULE

Several of the rules in this compilation involve a simple ratio which is of the order of 2/3. The ubiquity of the ratio suggests that it has some fundamental basis. Kanno (1981) has derived the 2/3 rule for the glass transition temperature,

$$T_g/T_m = 2/3$$

where T_g is the glass transition temperature and T_m is the melting point.

By analogy with the Lindemann equation. The success of this derivation suggests that the other 2/3 rules may have a similar origin.

3, RULE OF

In the equality of ratios, a:b = c:d, an "outside" unknown can be found by multiplying together the "inside unknowns" and dividing by the other "outside" unknown. Likewise, an "inside" unknown can be found by multiplying together the "outside unknowns" and dividing by the other "inside" unknown.

Here, the terms "inside" and "outside" refer to the imaginary space enclosed by the two colons. Thus, a and d are outside and b and c are inside. This very trivial rule dates back to the time before techniques of algebraic manipulation were familiar to the average artisan. It is included here for completeness and because it has a bearing on many other rules. That is, the numerous rules of Trouton's (qv) type involve a ratio which is equal to a constant. Rather than trying to memorize such constants, it is preferable to take two known items of data for a common material and a known property of the material in question. The unknown property for the target material can then be easily estimated by using the rule of 3. Although trivial, the rule has been considered in excruciating detail by Linderholm (1971). The latter author finds deep mathematical reasons for the non-existence of a "rule of one" or a "rule of two" and, indeed, for the impossibility of any even-numbered rule of this type. He goes on to find general expressions for rules of the form, $2n + 1$.

3 TEMPERATURES, RULE OF

See Prudhomme's rule.

4 VOLUMES, RULE OF

This rule (Porlezza, 1923) relates the critical volume, V_c, to the volume at the melting point, V_m, the volume at the boiling point, V_b, and the volume at absolute zero, V_o. Thus:

$$(V_o/V_c) + (V_m/V_c) + (V_b/V_c) = 1$$

4n + 2 RULE

See Hückel's rule.

6, RULE OF

In organic chemical reactions involving addition to an unsaturated function, the greater the number of atoms in the 6 position, the greater will be the steric hindrance to addition.

This rule was developed for the case of organic acids (Newman, 1950), and in this case the "6 position" is the sixth atom, counting from the carbonyl oxygen (Figure 58). The hindrance is reflected by their relative rates of esterification (Table 22). Really large steric effects are observed only in those acids having 9 or more atoms in the number 6 position. Although deduced on the basis of data for aliphatic compounds, the rule should be applicable to more complicated substances such as steroids.

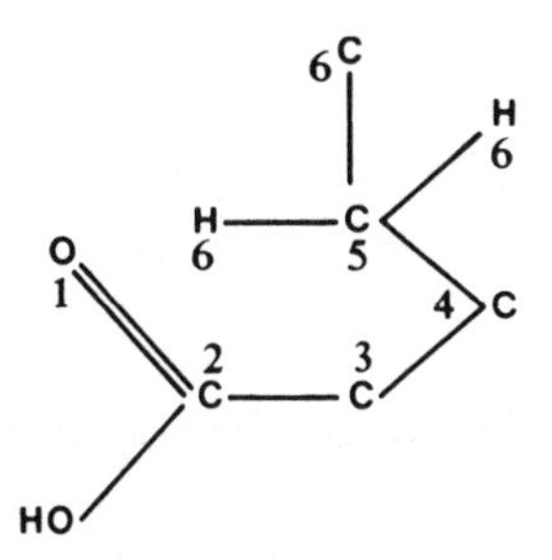

Figure 58. Illustrating the rule of 6. If one numbers the atoms in an organic acid by counting from the carbonyl O, it is seen that the feature which is present in butyric acid and higher homologues, but absent from lower homologues, is the presence of atoms in the number 6 position. The resultant steric hindrance affects the reaction rate.

Table 22
Rates of Esterification of Organic Acids
(relative to that of acetic acid)
See rule of 6

Acid	Atoms in Position 6	Relative Rate
acetic	0	1
propionic	0	0.69
isobutyric	0	0.20
butyric	3	0.34
isobutylacetic	3	0.31
methylethylacetic	3	0.05
methylpropylacetic	3	0.05
methylbutylacetic	3	0.04
methylisobutylacetic	3	0.04
isopropylacetic	6	0.07
methylisopropylacetic	6	0.01
diethylacetic	6	0.004
dipropylacetic	6	0.004

8, RULE OF

See the Octet rule.

8-N RULE

See the Hume-Rothery rules.

9, RULE OF

The method of checking arithmetical calculations by working in modulo 9 (also known as using "digital roots" or "casting out nines") is probably too well known to be worthy of inclusion here. However, it is often the belief that a rule is already known which impedes its wider dissemination (Stoneham, 1983b). So, briefly, in order to check an addition, say, one repeatedly adds the digits in order to obtain a single number and then checks the mathematical operation by performing the same calculation with the single digits ("digital roots"). Thus,

$$23786 + 55692 = 79478$$
$$8 + 9 = 8$$
$$8 = 8$$

One drawback of the method is that it cannot detect any error, such as the transposition of two adjacent digits, which introduces a discrepancy that is a multiple of nine. See the rule of 11.

11, RULE OF

This is similar to the rule of 9 but has the advantage that it will detect errors which arise from the transposition of two adjacent digits. It involves performing calculations with the remainder which result from dividing by 11. However, it is rather more tedious to divide by 11 than it is to add the digits; as in the rule of 9. On the other hand, this step can be speeded up by summing the odd digits (counting from the right) and subtracting the sum of the even digits. Thus, for 23786, $(6 + 7 + 2) - (3 + 8) = 4$.

13, RULE OF

A rule described by Bright and Chen (1983) for selecting possible empirical formulae corresponding to given mass spectra data. The first step is to assume that the substance is a hydrocarbon. The theoretical number of C and H atoms is then found by dividing the molecular weight, deduced from the mass spectrum, by 13. The hydrocarbon formula is immediately given by:

C_qH_{q+r}

where q is the quotient and r is the remainder. In addition, one can define a "degree of unsaturation," given by:

$u = (q - r + 2)/2$

This reflects the number of double bonds and/or rings present. For example, if the molecular weight is 78, the empirical formula resulting from the application of the rule is C_6H_6, with an unsaturation of 4, and benzene (1 ring + 3 double bonds) is immediately suspected. If no self-consistent hydrocarbon formula can be found, or if supplementary data indicate the presence of other elements, the method can be modified by balancing the mass and unsaturation via tentative substitutions (Table 23). Thus, the assumed presence of O can be balanced by subtracting a CH_4 "group," remembering that this also increases the unsaturation by unity. For example, if the mass is 142 and one O atom is assumed to be present, the formula deduced is $C_9H_{18}O$, with one double bond or ring. The substitution of N can be balanced by subtracting CH_2 and by increasing the unsaturation by 1/2, leading to the nitrogen rule (qv). The addition of ^{79}Br can be balanced by subtracting C_6H_7 or C_5H_{19}, with changes in unsaturation of − 3 and + 4, respectively. See also the Breakdown Rules.

Table 23
Parameters for Trial Substitutions
(according to the rule of 13)

Element	C/H equivalent	Change in Unsaturation
^{79}Br	C_6H_7	− 3
^{79}Br	C_5H_{19}	4
C	H_{12}	7
Cl	C_2H_{11}	3
F	CH_7	2
H_{12}	C	− 7
I	C_9H_{19}	0
I	$C_{10}H_7$	7
N	CH_2	0.5
N_2	C_2H_4	1
O	CH_4	1
P	C_2H_7	2
S^{2+}	C_2H_8	2
Si	C_2H_4	1

15% RULE

See the Hume-Rothery rules.

55 RULE

A suggested way (Ruekberg, 1987) of quickly spotting errors in elementary chemical calculations. It is based upon the fact that there are about 55 moles of water in a litre of pure water at normal temperature and pressure. As it would be difficult to get more solute into water than there is water in water, results indicating values greater than 55mole/l must be considered to be very dubious.

60% RULE

See Bastard's rule.

85% RULE

See Dingle's rule.

90-10 RULE

In order to find relevant data, 9 times as much additional irrelevant data will have to be brought into the main memory of a computer.

An empirically observed correlation (Banerjee et. al., 1978) concerning the management of large databases.

Appendices

Appendix 1
Melting Points, Boiling Points, and Critical Temperatures

Material	T_m(K)	T_b(K)	T_c(K)
Ag	1234	2436	4580
Ar	83.81	87.29	150.72
Au	1336	3081	5590
CCl_4	250.28	349.9	556.4
CH_4	90.68	111.66	191.1
CH_3Cl	175.43	248.93	416.2
Cd	594	1040	2900
Co	68.10	81.66	133.0
Cs	302	955	2043
Cu	1356	2839	5450
H_2	13.34	20.26	33.24
He	1.764	4.214	5.21
Hf	2500	4575	8110
Hg	234	630	1756
K	336	1031	2240
Kr	115.78	119.93	209.39
Li	453	1597	3223
Mo	2890	5100	9580
N_2	63.18	77.36	126.3
NH_3	195.4	239.73	405.51

continued

Appendix 1 (continued)
Melting Points, Boiling Points, and Critical Temperatures

Material	T_m(K)	T_b(K)	T_c(K)
Na	371	1156	2504
Nb	2740	5115	9880
Ne	24.544	27.15	44.44
Rb	327	967	2097
Rn	202	211	377.16
SO_2	197.69	263.13	430.27
V	2175	3682	6400
W	3680	5890	11810
Xe	161.36	165.03	289.75
Zn	693	1184	3620
Zr	2125	4682	8950

Appendix 2
Elastic Moduli of Metallic Elements

Metal	Structure	E(N/m²)	G(N/m²)	B(N/m²)
Ag	fcc	8.22×10^{10}	2.98×10^{10}	1.00×10^{11}
Al	fcc	7.74×10^{10}	2.61×10^{10}	7.89×10^{10}
Au	fcc	8.50×10^{10}	2.80×10^{10}	1.71×10^{11}
Ba	bcc	1.29×10^{10}	5.00×10^{9}	1.05×10^{10}
Ca	fcc	2.06×10^{10}	7.40×10^{9}	1.83×10^{10}
Cd	hcp	6.31×10^{10}	2.46×10^{10}	4.77×10^{10}
Ce	fcc	1.96×10^{10}	8.50×10^{9}	1.98×10^{10}
Co	hcp	2.16×10^{11}	8.21×10^{10}	1.67×10^{11}
Cr	bcc	2.40×10^{11}	9.00×10^{10}	1.97×10^{11}
Cs	bcc	1.79×10^{9}	6.60×10^{8}	8.01×10^{9}
Cu	fcc	1.26×10^{11}	4.77×10^{10}	1.34×10^{11}
Dy	hcp	6.31×10^{10}	2.48×10^{10}	4.33×10^{10}
Er	hcp	7.33×10^{10}	2.98×10^{10}	4.74×10^{10}
Eu	bcc	1.55×10^{10}	6.00×10^{9}	1.50×10^{10}
Fe	bcc	2.10×10^{11}	8.19×10^{10}	1.80×10^{11}
Fr	bcc	1.70×10^{9}	6.30×10^{8}	2.00×10^{9}
Gd	hcp	5.73×10^{10}	2.16×10^{10}	3.91×10^{10}
Hf	hcp	1.41×10^{11}	5.58×10^{10}	1.09×10^{11}

continued

Appendix 2 (continued)
Elastic Moduli of Metallic Elements

Metal	Structure	E(N/m²)	G(N/m²)	B(N/m²)
Ho	hcp	6.71×10^{10}	2.63×10^{10}	4.08×10^{10}
Ir	fcc	5.28×10^{11}	2.21×10^{11}	3.70×10^{11}
K	bcc	3.50×10^{9}	9.00×10^{8}	3.30×10^{9}
La	hcp	3.75×10^{10}	1.41×10^{10}	2.68×10^{10}
Li	bcc	1.17×10^{10}	4.30×10^{9}	1.18×10^{10}
Lu	hcp	6.85×10^{10}	2.72×10^{10}	4.76×10^{10}
Mg	hcp	4.52×10^{10}	1.73×10^{10}	3.45×10^{10}
Mo	bcc	3.13×10^{11}	1.18×10^{11}	2.85×10^{11}
Na	bcc	9.10×10^{9}	3.50×10^{9}	8.30×10^{9}
Nb	bcc	1.06×10^{11}	3.76×10^{10}	1.44×10^{11}
Nd	hcp	3.79×10^{10}	1.48×10^{10}	2.92×10^{10}
Ni	fcc	2.05×10^{11}	7.85×10^{10}	1.90×10^{11}
Os	hcp	5.50×10^{11}	2.14×10^{11}	4.26×10^{11}
Pb	fcc	2.35×10^{10}	8.60×10^{9}	4.04×10^{10}
Pd	fcc	1.37×10^{11}	5.21×10^{10}	1.84×10^{11}
Pm	hcp	4.30×10^{10}	1.70×10^{10}	3.60×10^{10}
Pr	hcp	3.52×10^{10}	1.38×10^{10}	2.55×10^{10}
Pt	fcc	1.90×10^{11}	6.37×10^{10}	2.75×10^{11}
Rb	bcc	2.35×10^{9}	6.30×10^{8}	2.30×10^{9}
Re	hcp	4.70×10^{11}	1.79×10^{11}	3.65×10^{11}
Rh	fcc	2.75×10^{11}	1.47×10^{11}	2.87×10^{11}
Ru	hcp	4.20×10^{11}	1.63×10^{11}	3.27×10^{11}
Sc	hcp	8.09×10^{10}	3.18×10^{10}	5.84×10^{10}
Sm	hcp	3.48×10^{10}	1.29×10^{10}	3.00×10^{10}
Sr	fcc	1.39×10^{10}	5.20×10^{9}	1.21×10^{10}
Ta	bcc	1.96×10^{11}	7.00×10^{10}	1.93×10^{11}
Tb	hcp	5.75×10^{10}	2.21×10^{10}	4.07×10^{10}
Tc	hcp	3.76×10^{11}	1.45×10^{11}	3.03×10^{11}
Th	fcc	7.34×10^{10}	2.84×10^{10}	5.77×10^{10}
Ti	hcp	1.08×10^{11}	4.34×10^{10}	8.86×10^{10}
Tm	hcp	7.40×10^{10}	2.91×10^{10}	4.62×10^{10}
V	bcc	1.31×10^{11}	4.81×10^{10}	1.39×10^{11}
W	bcc	3.94×10^{11}	1.51×10^{11}	3.38×10^{11}
Y	hcp	6.61×10^{10}	2.54×10^{10}	4.92×10^{10}
Yb	hcp	1.78×10^{10}	7.10×10^{9}	1.35×10^{10}
Zn	hcp	9.40×10^{10}	3.73×10^{10}	6.83×10^{10}
Zr	hcp	9.75×10^{10}	3.68×10^{10}	7.56×10^{10}

Appendix 3
Pauling Electronegativities of the Elements

Element	Electronegativity
Ac	1.1
Ag	1.9
Al	1.5
As	2.0
At	2.2
Au	2.4
B	2.0
Ba	0.9
Be	1.5
Bi	1.9
Br	2.8
C	2.5
Ca	1.0
Cd	1.7
Cl	3.0
Co	1.8
Cr	1.6
Cs	0.7
Cu	1.9
F	4.0
Fe	1.8
Fr	0.7
Ga	1.6
Ge	1.8
Hf	1.3
Hg	1.9
I	2.5
In	1.7
Ir	2.2
K	0.8
Li	1.0
Mg	1.2
Mn	1.5
Mo	1.8
N	3.0
Na	0.9
Nb	1.6

continued

Appendix 3 (continued)
Pauling Electronegativities of the Elements

Element	Electronegativity
Ni	1.8
O	3.5
Os	2.2
P	2.1
Pa	1.5
Pb	1.8
Pd	2.2
Po	2.0
Pt	2.2
Ra	0.9
Re	1.9
Rh	0.8
Ru	2.2
S	2.5
Sb	1.9
Sc	1.3
Se	2.4
Si	1.8
Sn	1.8
Sr	1.0
Ta	1.5
Tc	1.9
Te	2.1
Th	1.3
Ti	1.5
Tl	1.8
U	1.7
V	1.6
W	1.7
Y	1.2
Zn	1.6
Zr	1.4

References

Abegg, R. [1904]: *Zeitschrift für Anorganische Chemie,* 39, 330
Akhiezer, I. A., Davydov, L. N. [1981]: *Journal of Nuclear Materials*, 96, 115
Al-Bayyati, H. A. [1971]: *Technometrics,* 13, 675
Altona, C., Haasnoot, A. G. [1980]: *Organic Magnetic Resonance,* 13, 417
Anderson, P. W. [1963]: *Magnetism*, Academic Press, New York
Anderson, R. [1962]: *Solid State Electronics,* 5, 341
Angus, W. R., Hill, W. K. [1943]: *Transactions of the Faraday Society,* 39, 185
Antoncik, E. [1986]: *Radiation Effects,* 88, 217
Antonoff, A. [1907]: *Journal de Chimie Physique*, 5, 372
Argilés, J. M. [1986]: *Journal of Chemical Education*, 63, 927
Ascarelli, P. [1971]: Technical Report 71-47, US Army Materials and Mechanics Research Center
Babinet, G. [1838]: *Comptes Rendues,* 7, 832
Badger, R. M. [1934]: *Journal of Chemical Physics*, 2, 128
Badger, R. M. [1935]: *Journal of Chemical Physics*, 3, 710
Balaban, A. T. [1981]: *Revue Roumaine de Chimie*, 26, 407
Bancroft, A. M. [1981]: *The Physics Teacher,* December, p.615
Banerjee, J., Baum, R., Hsiao, D. K. [1978]: ACM *Transactions on Database Systems,* 3–4, 347
Bangwei, Z. [1983]: *Physica*, 121B, 405
Barat, M., Lichten, W. [1972]: *Physical Review*, A6, 211
Barber, N. F. [1932]: *Proceedings of the Leeds Philosophical Society,* 2, 427
Barnsley, M. F., Ervin, V., Hardin, D., Lancaster, J. [1986]: *Proceedings of the National Academy of Sciences of the USA,* 83, 1975

Bartenev, G. M., Sanditov, D. S. [1982]: *Journal of Non-Crystalline Solids*, 48, 405
Bastard, G. [1981]: *Physical Review B*, 24, 5693
Beaman, R. G. [1952]: *Journal of Polymer Science*, 9, 470
Beke, D. L., Kedves, F. J. [1983]: *Zeitschrift für Metallkunde,* 74, 238
Benford, F. [1938]: *Proceedings of the American Philosophical Society,* 78, 551
Bénière, F. [1974]: *Journal de Physique* - Lettres, 35, L9
Benson, S. W. [1976]: *Thermochemical Kinetics*, Wiley-Interscience, New York
Bent, A. R. [1965]: *The Second Law*, Oxford, New York, p.271
Bentley, J. [1985]: Communications of the ACM, 28, 896
Berezin, A. A. [1972]: *Physica Status Solidi,* 49b, 51
Berezin, A. A. [1976]: *Fizika Tverdogo Tela,* 18, 493
Berezin, A. A. [1979]: *Journal of Physics* C, 12, L363
Berezin, V. M., Vyatkin, G. P. [1986]: *Fizika Tverdogo Tela,* 28, 1177
Bernal, J. D., Fowler, R. [1933]: *Journal of Chemical Physics*, 1, 515
Beynon, J. H. [1960]: *Mass Spectrometry and its Applications to Organic Chemistry*, Elsevier, p.307
Bingham, C. [1906]: *Journal of the American Chemical Society*, 28, 723
Blanck, H. F. [1989]: *Journal of Chemical Education*, 66, 757
Bloom, E. G., Mohler, F. L., Lengel, J. H., Wise, C. E. [1948]: *Journal of Research of the National Bureau of Standards*, 41, 129
Bogdan, K. [1913]: *Zeitschrift für Physikalische Chemie*, 82, 93
Bollmann, W. [1980]: *Physica Status Solidi*, 61a, 395
Bonzel, H. P. [1972]: *Structure and Properties of Metal Surfaces* (edited by S. Shimodaira), Maruzen, Tokyo
Boyer, R. F. [1954]: *Journal of Applied Physics*, 25, 825
Brauer, H., Kriegel, E. [1965]: *Chemie-Ingenieur-Technik,* 37, 265
Brewer, L. [1932]: in *Electronic Structure and Alloy Chemistry of the Transition Elements* (edited by P. A. Beck), Wiley, New York
Brewster, J. H. [1959]: *Journal of the American Chemical Society*, 81, 5475
Bright, J. W., Chen, E. C. M. [1983]: *Journal of Chemical Education*, 60, 557
Brown, N. [1967]: in *Intermetallic Compounds* (edited by J. H. Westbrook), Wiley, New York, p.272
Bruhat, D. [1915]: *Annales de Physique*, 3, 232
Burden, B. [1871]: *Philosophical Magazine*, 41, 528
Bussemer, P. [1979]: *Physica Status Solidi*, 94b, K77

Carnelly, R. [1878]: *Berichte*, 11, 2289
Carpenter, A. K. [1983]: *Journal of Chemical Education*, 60, 562
Carpenter, B. K. [1984]: *Determination of Organic Reaction Mechanisms* Wiley, New York, p.221

Cartan, L. [1937]: *Journal de Physique*, 11, 453
Caspers, L. M., Van Veen, A. [1981]: *Physica Status Solidi*, 68a, 339
Catchpole, A. G., Hughes, E. D., Ingold, C. K. [1948]: *Journal of the Chemical Society*, 11
Chapnik, I. M. [1980]: *Journal of Materials Science*, 15, 3175
Chapnik, I. M. [1984a]: *Physica Status Solidi*, 123b, K183
Chapnik, I. M. [1984b]: *Journal of Physics* F, 14, 1919
Chapnik, I. M. [1984c]: *Physica Status Solidi*, 125b, K189
Chapnik, I. M. [1985]: *Journal of Materials Science Letters*, 4, 370
Chapnik, I. M. [1986]: *Physica Status Solidi*, 137b, K95
Charney, E. [1965]: *Tetrahedron*, 21, 3127
Cho, S. A. [1977]: *Acta Metallurgica*, 25, 1085
Chwolson, C. [1923]: *Lehrbuch der Physik*, Brunswick, 3, 210
Clemens, B. M. [1986]: *Physical Review* B, 33, 7615
Clough, G. W. [1918]: *Journal of the Chemical Society*, 113, 526
Colson, F. [1887]: *Comptes Rendues*, 104, 428
Condon, F. E. [1954]: *Journal of Chemical Education*, 31, 651
Couchman, P. R. [1975]: *Physics Letters*, 54A, 309
Coulson, C. A., Rushbrooke, G. S. [1940]: *Proceedings of the Cambridge Philosophical Society*, 36, 193
Cram, D. J., Abd Elhafez, F. A. [1952]: *Journal of the American Chemical Society*, 74, 5828
Darzens, F. [1897]: *Comptes Rendues*, 124, 610
De Forcrand, R. [1901]: *Comptes Rendues*, 132, 879
De Mallemann, A. [1923]:, *Journal de Physique*, 4, 29
Dingle, R. [1975]: *Festkörperprobleme*, 15, 21
Do, Y. J., Yen, T. T., Chen, L. R. [1984]: *Journal of Physics* F, 14, L141
Doherty, R. D. [1978]: *Scripta Metallurgica*, 12, 591
Donald, J. W., Davies, H. A. [1978]: *Journal of Non-Crystalline Solids*, 30, 77
Eberhart, J. G. [1966]: *Transactions of the Metallurgical Society of AIME*, 236, 1362
Eckroth, D. R. [1967]: *Journal of Organic Chemistry*, 32, 3362
Eggert, F. [1932]: *Physical Chemistry*, 236
Eichler, J., Wille, U. [1974]: *Physical Review Letters*, 33, 56
Eliel, E. L., Allinger, N. L., Angyall, S. J., Morrison, G. A. [1967]: *Conformational Analysis*, Interscience
Fajans, K. [1931]: *Radio-elements and Isotopes, Chemical Forces and Optical Properties of Substances*, McGraw-Hill, New York
Farlow, G. C., Appleton, B. R., Boatner, L. A., McHargue, C. J., White, C. W. , Clark, G. J., Baglin, J. E. E. [1985]: *Proceedings of the Materials Research Society Symposium*, 45, 127
Filonenko, V. A. [1970]: *Russian Journal of Physical Chemistry*, 44, 648

Findlay, A. [1951]: *'The Phase Rule'*, Dover, New York, p.353
Fine, M. E., Brown, L. D., Marcus, H. L. [1984]: *Scripta Metallurgica*, 18, 951
Finnie, I., Wolak, J., Kabil, Y. [1967]: *Journal of Materials*, 2, 683
Finucan, H. M. [1973]: *Mathematical Gazette*, 57, 203
Flynn, C. P. [1988]: *Journal of Physics F*, 18, L195
Frankland, G. [1899]: *Journal of the Chemical Society*, 75, 347
Freudenberg, K., Kuhn, W., Bumann, I. [1930]: *Berichte*, 63, 2380
Friederich, E. [1925]: *Zeitschrift für Physik*, 31, 813
Fuchs, G. E., Ficalora, P. J. [1985]: *Scripta Metallurgica*, 19, 275
Furdyna, J. K., Kossut, J. [1986]: *Superlattices and Microstructures*, 2, 89

Giambiagi, M. S., Giambiagi, M. [1979]: *Lettere al Nuovo Cimento*, 25, 459
Gilbert, H. F. [1977]: *Journal of Chemical Education*, 54, 492
Goodstein, M. P. [1970]: *Journal of Chemical Education*, 47, 452
Gorecki, T. [1974]: *Zeitschrift für Metallkunde*, 65, 426
Gorecki, T. [1979]: *Zeitschrift für Metallkunde*, 70, 121
Gorecki, T. [1980]: *Materials Science and Engineering*, 43, 225
Goryunova, N. A. [1965]: *Chemistry of Diamond-Like Semiconductors*, Chapman & Hall, London
Goudsmit, S. A. [1928]: *Physical Review*, 31, 946
Gould, E. S. [1955]: *Inorganic Reactions and Structure*, Holt, Rinehart & Winston, New York
Groshans, K. [1847]: *Annalen der Physik*, 78, 112
Grunberg, A., Nissan, B. [1948]: *Nature*, 161, 170
Guldberg, C. M. [1890]: *Zeitschrift für Physikalische Chemie*, 5, 374
Gupta, H., Morral, J. E., Nowotny, H. [1986]: *Scripta Metallurgica*, 20, 889
Gust, W., Mayer, S., Bögel, A., Predel, B. [1985]: *Journal de Physique-Colloque C4*, 46, 537
Gutmann, V. [1978]: *The Donor-Acceptor Approach to Molecular Interactions*, Plenum, New York

Hägg, G. [1930]: *Zeitschrift für Physikalische Chemie*, B6, 221
Hammett, L. P. [1937]: *Journal of the American Chemical Society*, 59, 96
Hanack, M. [1965]: *Conformational Theory*, Academic Press
Hardy, W. B. [1900]: *Proceedings of the Royal Society*, 66, 110
Hauffe, K. [1951]: *Zeitschrift für Metallkunde*, 42, 34
Hermann, H. [1931]: *Zeitschrift für Anorganische Chemie*, 198, 204
Herz, G. [1920]: *Zeitschrift für Anorganische Chemie*, 112, 278
Herz, G. [1929a]: *Zeitschrift für Anorganische Chemie*, 179, 277
Herz, G. [1929b]: *Zeitschrift für Anorganische Chemie*, 180, 284
Hillert, M. [1985]: *International Metals Reviews*, 30, 45

Hillert, M. [1988]: *Scripta Metallurgica*, 22, 1085
Holleran, E. M., Jespersen, N. D. [1980]: *Journal of Chemical Education*, 57, 670
Hondros, E. D. [1980]: *Scripta Metallurgica*, 14, 345
Houghton, E. L., Boswell, R. P. [1969]: *Further Aerodynamics for Engineering Students*, Edward Arnold, p.467
Hovland, A. K. [1986]: *Journal of Chemical Education*, 63, 607
Hsu, T. Y. [1985]: *Journal of Materials Science*, 20, 23
Hudson, C. S. [1909]: *Journal of the American Chemical Society*, 31, 66
Hudson, C. S. [1910]: *Journal of the American Chemical Society*, 32, 338
Hudson, C. S. [1939]: *Journal of the American Chemical Society*, 61, 1525
Hund, F. [1925]: *Zeitschrift für Physik*, 33, 345
Hutchings, I. M. [1975]: *Wear*, 35, 371

Ingold, C. K. [1953]: *Structure and Mechanism in Organic Chemistry*, Cornell University Press, New York, p.933

Kagawa, A., Okamoto, T., Saito, K., Ohta, M. [1984]: *Journal of Materials Science*, 19, 2546
Kallina, C., Lynn, J. [1976]: *Interfaces*, 7, 37
Kanno, H. [1981]: *Journal of Non-Crystalline Solids*, 44, 409
Kapellos, S., Mavrides, A. [1987]: *Journal of Chemical Education*, 64, 941
Karplus, M. [1959]: *Journal of Chemical Physics*, 30, 11
Katz, R. N. [1985]: *Materials Science and Engineering*, 71, 227
Kauffman, J. M. [1986]: *Journal of Chemical Education*, 63, 474
Kempster, C. J. E., Lipson, H. [1972]: *Acta Crystallographica*, B28, 3674
Kereselidze, T. M., Kikiani, B. I. [1984]: *Soviet Physics - JETP*, 60, 423
Kern, D. Q. [1950]: *Process Heat Transfer*, McGraw-Hill, New York
Keyes, R. W. [1962]: *Journal of Applied Physics*, 33, 3371
Khruschov, M. M. [1974]: *Wear*, 28, 69
Kireev, K. [1929]: *Journal of the Russian Physical Chemistry Society*, 61, 1369
Koch, J. M., Koenig, C. [1986]: *Philosophical Magazine* B, 54, 177
Kolb, D. [1978]: *Journal of Chemical Education*, 55, 326
Koo, Y. M., Cohen, J. B. [1987]: *Materials Science and Engineering*, 91, L5
Kopp, G. [1842]: *Annalen*, 41, 79
Kramer, H. L., Herschbach, D. R. [1970]: *Journal of Chemical Physics*, 53, 2792
Kubaschewski, O. [1984]: *High Temperatures-High Pressures*, 16, 197
Kuhlmann, K. [1914]: *Archiv für Electrotechnik*, 3, 203
Kumler, R. [1935]: *Journal of the American Chemical Society*, 57, 600
Küster, S. [1890]: *Zeitschrift für Physikalische Chemie*, 5, 322

Lapworth, R. [1898]: *Journal of the Chemical Society*, 73, 445
Laugier, M. T. [1985]: *Journal of Materials Science Letters*, 4, 211

Laves, F. [1967]: in *Intermetallic Compounds* (editor J. H. Westbrook), Wiley, p.131

Lazaridou, M., Varotsos, C., Alexopoulos, K., Varotsos, P. [1985]: *Journal of Physics C*, 18, 3891

Le Claire, A. D. [1976]: Proceedings of the 19th Metallurgical Colloquium, Saclay, France

Lee, H. C., Kim, Y. S. [1981]: *Journal of Chemical Physics*, 74, 6144

Lee, T. C., Robertson, I. M., Birnbaum, H. K. [1989]: *Scripta Metallurgica*, 23, 799

Lemieux, R. U. [1971]: *Pure and Applied Chemistry*, 25, 526

Levene, P. A. [1915]: *Journal of Biological Chemistry Chem.*, 23 145

Linderholm, C. E. [1971]: *Mathematics Made Difficult*, Wolfe, London, p.143

Lister, D. G., MacDonald, J. N., Owen, N. L. [1978]: *Internal Rotation and Inversion*, Academic Press

Liu, B. X., Johnson, W. L., Nicolet, M. A., Lau, S. S. [1983]: *Applied Physics Letters*, 42, 45

Livingstone, D. [1936]: *Science*, 84, 459

Lobo, V. M. M., Mills, R. [1982]: *Electrochimica Acta*, 7, 969

Longchambon, G. [1921]: *Comptes Rendues*, 172, 1187

Longinescu, G. [1903]: *Journal de Chimie Physique*, 1, 296

Lorenz, K. [1916]: *Zeitschrift für Anorganische Chemie*, 94, 240

Lutz, O., Jirgensons, B. [1930]: *Berichte*, 63, 448

Lutz, O., Jirgensons, B. [1931]: *Berichte*, 64, 1221

Luzzi, D. E., Meshii, M. [1986]: *Scripta Metallurgica*, 20, 943

MacKenzie, I. K., Lichtenberger, P. C. [1976]: *Applied Physics*, 9, 331

March, N. H., Richardson, D. D., Tosi, M. P. [1980]: *Solid State Communications*, 35, 903

Markownikoff, V. V. [1870]: *Annalen*, 153, 256

Markownikoff, V. V. [1875]: *Comptes Rendues*, 81, 670

Matthias, B. T. [1957]: in *Progress in Low-Temperature Physics*-Volume 2 (edited by C. J. Gorter), North-Holland, Amsterdam, p.138

Mayer, F. [1928]: *Zeitschrift für Physik*, 85, 278

McAdam, G. D. [1951]: *Journal of the Iron and Steel Institute*, 168, 348

McCaldin, J. O., McGill, T. C., Mead, C. A. [1976]: *Physical Review Letters*, 36, 56

Meislich, H. [1963]: *Journal of Chemical Education*, 40, 401

Meyerhoffer, W. [1904]: *Berichte*, 37, 2604

Michael, A. [1899]: *Journal für Praktische Chemie*, 60, 291

Mock, R., Güntherodt, G. [1984]: *Journal of Physics C*, 17, 1984

Mogro-Campero, A. [1982]: *Journal of Applied Physics*, 53, 1224

Montgomery, G. [1911]: *American Chemical Journal*, 46, 298
Moore, M. A. [1974]: PhD Thesis, University of Newcastle
Mooser, E., Pearson, W. B. [1956]: *Journal of Electronics*, 1, 629
Moss, T. S. [1950]: *Proceedings of the Physical Society,* B63, 167
Mott, B. W. [1956]: *Micro-indentation Hardness Testing*, London
Mowery, D. F. [1969]: *Journal of Chemical Education*, 46, 269
Mukherjee, K. [1964]: *Physics Letters*, 8, 17
Myers, R. T. [1981]: *Journal of Chemical Education*, 58, 681
Nabarro, F. R. N. [1988]: *Philosophical Magazine A*, 57, 565
Nakamura, T. [1981]: *Japanese Journal of Applied Physics*, 20, L653
Negita, K. [1985]: *Journal of Materials Science Letters*, 4, 417
Newman, M. S. [1950]: *Journal of the American Chemical Society*, 72, 4783
Newton, J. [1963]: *Extractive Metallurgy*, Wiley, New York
Osenbach, J. W., Bitler, W. R., Stubican, V. S. [1981]: *Journal of the Physics and Chemistry of Solids*, 42, 599
Ostwald, W. [1897]: *Zeitschrift für Physikalische Chemie*, 22, 306
Otsuka, S., Kozuka, Z., Chang, Y. A. [1984]: *Metallurgical Transactions*, 15B, 329
Owen, R. P. [1968]: *Electronics*, 41 (19th August), 87
Pak, H. R., Inal, O. T. [1987]: *Journal of Materials Science*, 22, 1945
Pamplin, B. R. [1964]: *Journal of the Physics and Chemistry of Solids*, 25, 675
Patil, R. V., Tiwari, G. P., Sharma, B. D. [1980]: *Metal Science*, 14, 525
Patterson, J. W., Brode, W. R. [1943]: *Archives of Biochemistry*, 2, 247
Pauli, W. [1923]: *Zeitschrift für Physik*, 16, 155
Pelton, A. D. [1988]: *Metallurgical Transactions*, 1988, 19A, 1819
Pfeiffer, P., Christeleit, W. [1937]: *Zeitschrift für Physiologische Chemie*, 245, 197
Pictet, R. [1876]: *Philosophical Magazine*, 1, 477
Pilling, N. B., Bedworth, R. E. [1923]: *Journal of the Institute of Metals*, 29, 529
Poritsky, H. [1938]: *Transactions of the AIEE,* 57, 727
Porlezza, R. [1923]: *Nuovo Cimento*, 25, 291, 305
Prelog, V. [1964]: *Pure and Applied Chemistry*, 9, 119
Prudhomme, G. [1920]: *Journal de Chimie Physique*, 18, 270
Pugh, S. F. [1954]: *Philosophical Magazine*, 45, 823
Pulfrich, R. [1889]: *Zeitschrift für Physikalische Chemie*, 4, 561
Ravindra, N. M., Srivastava, V. K. [1979]: *Infrared Physics*, 19, 603
Reddy, P. A. [1987]: *Journal of Chemical Education*, 64, 400
Reid, K. F. [1968]: *Properties and Reactions of Bonds in Organic Molecules*, Longmans, p.154
Remick, A. E. [1949]: *Electronic Interpretations of Organic Chemistry*, Wiley, New York, p.211
Reynolds, C. L., Couchman, P. R. [1974]: *Physics Letters*, 50A, 157

Reynolds, C. L., Couchman, P. R., Karasz, F. E. [1976]: *Philosophical Magazine*, 34, 659
Rice, J. R., Thomson, R., [1974]: *Philosophical Magazine*, 29, 73
Richardson, F. D. [1950]: *Journal of the Iron and Steel Institute*, 166, 137
Richardson, L. F. [1908]: *Philosophical Magazine*, 15, 237
Rouvray, D. H. [1986]: *Scientific American*, 255[3], 36
Roux, F., Vignes, A. [1970]: *Revue de Physique Appliquée*, 5, 393
Ruekberg, B. P. [1987]: *Journal of Chemical Education*, 64, 892
Sangwal, K. [1982]: *Crystal Research and Technology*, 17, K21
Savitsky, G. B. [1960]: *American Journal of Physics*, 28, 12
Saytzeff, A. [1875]: *Annalen*, 179, 296
Schäfer, H. [1964]: *Chemical Transport Reactions*, Academic Press, p.31
Schmidt, O. [1934]: *Berichte*, 67, 62
Schreinemakers, F. A. H. [1915]: *Proceedings of the Akademie van Wetenschappen*, 18, 116
Schulze, H. [1882]: *Journal für Praktische Chemie*, 25, 431
Schulze, H. [1883]: *Journal für Praktische Chemie*, 27, 320
Sher, A., Chen, A. B., Spicer, W. E. [1985]: *Applied Physics Letters*, 46, 54
Sherby, O. D., Simnad, M. T. [1961]: *Transactions of the ASM*, 54, 227
Shewmon, P. G. [1963]: *Diffusion in Solids*, McGraw-Hill, New York
Shinskey, G. [1971]: *Instruments and Control Systems*, 44 (December), 11
Shirley, A. I., Hall, C. K. [1984]: *Acta Metallurgica*, 32, 49
Shukla, M. M. [1982]: *Journal of Physics D*, 15, L177
Sigarev, S. E., Galiulin, R. V. [1985]: *Kristallografiya*, 30, 1013
Skita, A. [1923]: *Berichte*, 79, 3443
Smeltzer, C. E., Gulden, M. E., McElmury, S. S., Compton, W. A. [1970]: Report 70-36, US Army Aviation Materials Laboratory
Smith, F. T. [1972]: *Physical Review A*, 5, 1708
Soliman, M. A., Kinawy, N., Eleiwa, M. M. [1986]: *Journal of Materials Science Letters*, 5, 329
Somorjai, G. A., Szalkowski, F. J. [1971]: *Journal of Chemical Physics*, 54, 389
Sottini, S., Russo, V., Righini, G. C. [1979]: *IEEE Transactions on Circuits and Systems*, 26, 1036
Stalker, M. K., Morral, J. E. [1988]: *Scripta Metallurgica*, 22, 1787
Stevenson, D. P. [1951]: *Discussions of the Faraday Society*, 10, 35
Stohmann, F. [1889]: *Journal für Praktische Chemie*, 40, 357
Stone, H. E. N. [1984]: *Journal of Materials Science Letters*, 3, 807
Stoneham, A. M. [1983a]: *Applications of Surface Science*
Stoneham, A. M. [1983b]: personal communication
Streeter, V. L. [1962]: *Fluid Mechanics*, McGraw-Hill, p.330
Stwalley, W. C. [1971]: *Journal of Chemical Physics*, 55, 170
Sunberg, R. J. [1986]: *Journal of Chemical Education*, 63, 714

Swain, C. G., Kuhn, D. A., Schowen, R. L. [1965]: *Journal of the American Chemical Society*, 87, 1553
Swain, C. G., Thornton, E. R. [1962]: *Journal of the American Chemical Society*, 81, 817

Tabor, D. [1970]: *Reviews of Physics in Technology*, 1, 145
Taft, R. W. [1953]: *Journal of the American Chemical Society*, 75, 4534
Thole, B. T., Van der Laan, G. [1987]: *Europhysics Letters*, 4, 1083
Tiwari, G. P. [1978]: *Transactions of the Japan Institute of Metals,* 19, 125
Trahanovsky, W. S. [1971]: *Functional Groups in Organic Chemistry*, Prentice-Hall, New York, p.17
Traube, F. [1891]: *Ann.,* 265, 27
Traube, F. [1899]: *Sammlung Chemischer und Chemisch-Technischer Vorträge,* 4, 255
Trouton, F. [1884]: *Philosophical Magazine*, 18, 54
Tschugaeff, L. [1898]: *Berichte*, 31, 360
Urbach, F. [1953]: *Physical Review*, 92, 1324
Uvarov, N. F., Hairetdinov, E. F., Boldyrev, V. V. [1984]: *Journal of Solid State Chemistry*, 51, 59
Van Arkel, A. E. [1932]: *Recueil des Travaux Chimiques des Pays-Bas,* 51, 1081
Van Dyck, D., Colaitis, D., Amelinckx, S. [1981]: *Physica Status Solidi*, 68a, 385
Van Liempt, J. [1935]: *Zeitschrift für Physik*, 96, 534
Van Uitert, L. G. [1979]: *Journal of Applied Physics*, 50, 8052
Van Uitert, L. G. [1981]: *Journal of Applied Physics*, 52, 5553
Van't Hoff, F. H. [1898]: *The Arrangement of Atoms in Space*, London
Varotsos, P., Alexopoulos, K., Lazaridou, M. [1984]: *Physica Status Solidi*, b125, K109
Vijh, A. K. [1975]: *Journal of Materials Science*, 10, 998
Von Auwers, K. [1920]: *Annalen*, 420, 84
Von Barth, U., Grossmann, G. [1979]: *Solid State Communications*, 32, 645
Von Eötvös, R. [1886]: *Annalen der Physik*, 27, 452
Vorlander, G. [1919]: *Berichte*, 52, 263
Voronel, A., Rabinovich, S. [1987]: *Journal of Physics F*, 17, L193
Warren, B. E. [1933]: *Zeitschrift für Kristallographie*, 86, 349
Watson, K. M. [1931]: *Industrial and Engineering Chemistry*, 23, 360
Weertman, J. [1981]: *Philosophical Magazine A*, 43, 1103
Werner, A. [1912]: *Bulletin de la Société Chimique de France*, 11, 1
Wilson, A. J. C. [1944]: *Journal of the Institute of Metals*, 70, 543
Winther, C. [1895]: *Berichte*, 28, 3000
Wolf, W. [1965]: *Tetrahedron Letters*, 1075

Woodsum, H. C. [1978]: *American Journal of Physics*, 46, 770
Yavari, A. R., Rouault, A. [1986]: *Journal of Physics F*, 16, 687
Yi, P. J. [1947]: *Journal of Chemical Education*, 24, 567
Yoshida, A. [1984]: Transactions of the Japan Institute of Metals, 25, 105
Zachariasen, W. J. [1932]: *Journal of the American Chemical Society*, 54, 3841
Zehe, A., Röpke, G. [1986]: *Journal of Physics F*, 16, 407
Zener, C. [1951]: *Journal of Applied Physics*, 22, 372
Ziemann, P. [1985]: *Materials Science and Engineering*, 69, 95
Zintl, E., Goubeau, J., Dullenkopf, W. [1931]: *Zeitschrift für Physikalische Chemie*, A154, 1

General Subject Index

Dependent Variable Index

* Asterisks indicate rules that have not been "formally" adopted. See Introduction for explanation.

*Asterisks indicate rules that have not been "formally" adopted. See Introduction for explanation.

*Asterisks indicate rules that have not been "formally" adopted. See Introduction for explanation.

*Asterisks indicate rules that have not been "formally" adopted. See Introduction for explanation.

*Asterisks indicate rules that have not been "formally" adopted. See Introduction for explanation.

*Asterisks indicate rules that have not been "formally" adopted. See Introduction for explanation.

*Asterisks indicate rules that have not been "formally" adopted. See Introduction for explanation.

Independent Variable Index

*Asterisks indicate rules that have not been "formally" adopted. See Introduction for explanation.

* Asterisks indicate rules that have not been "formally" adopted. See Introduction for explanation.

* Asterisks indicate rules that have not been "formally" adopted. See Introduction for explanation.

* Asterisks indicate rules that have not been "formally" adopted. See Introduction for explanation.

* Asterisks indicate rules that have not been "formally" adopted. See Introduction for explanation.